6318198

Research Methods for Construction

D1537005

Research Methods for Construction

Richard Fellows
*School of Architecture & Civil Engineering,
University of Bath*

and

Anita Liu
*Department of Real Estate and Construction,
University of Hong Kong*

**Blackwell
Science**

© 1997 by
Blackwell Science Ltd
Editorial Offices:
Osney Mead, Oxford OX2 0EL
25 John Street, London WC1N 2BL
23 Ainslie Place, Edinburgh EH3 6AJ
350 Main Street, Malden
  MA 02148 5018, USA
54 University Street, Carlton
  Victoria 3053, Australia

Other Editorial Offices:

Blackwell Wissenschafts-Verlag GmbH
Kurfürstendamm 57
10707 Berlin, Germany

Blackwell Science KK
MG Kodenmacho Building
7-10 Kodenmacho Nihombashi
Chuo-ku, Tokyo 104, Japan

First published 1997

Set in 11 on 14pt Bembo
by Aarontype Ltd, Bristol, England
Printed and bound in Great Britain
by Hartnolls Ltd, Bodmin,
Cornwall

The Blackwell Science logo is a trade mark
of Blackwell Science Ltd, registered at the
United Kingdom Trade Marks Registry

DISTRIBUTORS

Marston Book Services Ltd
PO Box 269
Abingdon
Oxon OX14 4YN
(*Orders*: Tel: 01235 465500
             Fax: 01235 465555)

USA
  Blackwell Science, Inc.
  Commerce Place
  350 Main Street
  Malden, MA 02148 5018
  (*Orders*: Tel: 800 759 6102
                  617 388 8250
             Fax: 617 388 8255)

Canada
  Copp Clark Professional
  200 Adelaide Street West, 3rd Floor
  Toronto, Ontario M5H 1W7
  (*Orders*: Tel: 416 597-1616
                  800 815-9417
             Fax: 416 597-1617)

Australia
  Blackwell Science Pty Ltd
  54 University Street
  Carlton, Victoria 3053
  (*Orders*: Tel: 03 9347-0300
             Fax: 03 9347-5001)

A catalogue record for this title
is available from the British Library

ISBN 0-632-04244-3

Library of Congress
Cataloging-in-Publication Data

Fellows, Richard
    Research methods for construction/Richard
    Fellows and Anita Liu
        p.   cm.
    Includes bibliographical references and index
    ISBN 0-632-04244-3 (pb)
    1. Building-Research-Methodology.
    I. Liu, Anita. II. Title
  TH213.5.F45 1997   97-21932
  624'.072-dc21   CIP

Contents

Preface

'The scientist is not the person who knows a lot but rather the person who is not prepared to give up the search for truth.'

Popper, 1989, p. 334; reporting Marx and Engels.

A discipline or profession is established by developing a body of knowledge which is unique – that body of knowledge is produced through research. Construction draws on a wide variety of established subjects, including natural sciences, social sciences, engineering and management, and applies them to its particular context and requirements. Only by use of appropriate methodologies and methods of research, applied with rigour, can the body of knowledge for construction be established and advanced with confidence.

Although a number of texts are available discussing research methodologies and methods generally, there is a notable lack of such books in construction. Statistics, philosophy, natural and social sciences have produced relevant texts; this book is aimed at the broad discipline of construction. In particular, the contents of this book will be useful to students of building, civil engineering, architecture, construction management and all forms of surveying, whether researching for dissertations for Bachelors or Masters degrees or undertaking research for Masters degrees or Doctorates. Further, the book will be helpful to practitioners and students in these disciplines in providing guidance on how to instil rigour in problem-solving and on producing reports and publications.

The approach adopted in the book is to outline the process of research: the initial recognition that research is necessary; the development of a proposal; the execution of the research; the drawing of conclusions; and the production and presentation of the final report. The book comprises three main sections – producing a proposal, executing the research and reporting the results. The book discusses the main issues in research and examines the primary approaches – both qualitative and quantitative. The methods adopted for scientific and engineering experiments and simulations are evaluated as well as those employed for research into managerial issues, and social and economic investigations.

In considering the requirements for data and data analyses, the book presents discussion of important statistical considerations and techniques. These enable the researcher to appreciate the issues which need to be evaluated in devising how research may be carried out effectively and efficiently in the practical environment of modern construction activity. Thus, the book considers a range of methodologies and methods to facilitate selection of the most appropriate research approach to adopt (from an informed perspective). It provides sufficient depth in examination of the subject materials to facilitate the execution of research projects.

Increasingly, pleas are voiced seeking special treatment (leniency) for research in construction, based on arguments concerning the particular nature of construction and the problems of research which ensue. Unless the research proposed and undertaken in construction can withstand scrutiny on the same bases as all other research, the discipline will fail to advance adequately. As construction is of major importance to all societies and economies, it is essential that the discipline advances as rapidly and as rigorously as possible.

This book results from the combined experiences of the authors in executing, supervising and managing many types of research projects over a number of years, in UK and Hong Kong in particular. In producing this book, we seek to provide insights to ease the, often, novel and difficult path through the research process so that common problems may be overcome (if not avoided), progress can be efficient, effective and rigorous and, perhaps most important, enjoyable and rewarding.

We are immensely grateful to the many colleague-friends who have provided much encouragement and help, to the anonymous referees for their constructive and valuable recommendations on contents and style and, especially, to Julia Burden at Blackwell Science for her invaluable assistance, tolerance and patience in guiding us through the tortuous path of producing this book. Responsibility for the contents, any errors and contentious issues, remains ours alone.

Part 1

Producing a Proposal

Chapter 1

Introduction

The objectives of this chapter are to:

- introduce the **concept of research**;
- provide awareness of different **classifications of research**;
- outline the essentials of **theories and paradigms**;
- discuss the various **research styles**;
- introduce **quantitative and qualitative approaches**;
- consider **where to begin.**

The concept of research

Chambers English Dictionary defines research as:

- a careful search
- investigation
- systematic investigation towards increasing the sum of knowledge.

For many people, the prospect of embarking on a research project is a daunting one. However, especially for people who are associated with a project-oriented industry, such as property development, building design, construction, or facilities management, familiarity with the nature of projects and their management is a significant

3

advantage. Dr Martin Barnes, an ex-chairperson of the Association of Project Managers (APM), has described a project as a task or an activity which has a beginning (start), a middle and an end. Despite the situation that much research is carried out as part of a long term 'rolling' programme, each individual package of research is itself a project – an entity which is complete in itself, whilst contributing to the overall programme.

Indeed, any work which assists in the advancement of knowledge, whether of society, a group or an individual, involves research; it will involve enquiry and learning also.

Research: a careful search/investigation

Research can be considered to be a 'voyage of discovery', whether anything is discovered or not. What is discovered depends on the pattern and techniques of searching, the location and subject material investigated and the analyses carried out. The knowledge and abilities of researchers and their associates are important in executing the investigative work and, perhaps more especially, in the production of results and the drawing of conclusions.

Research: contribution to knowledge

Traditionally, the essential feature of research for a doctoral degree (PhD) is that the work makes an original (incremental) contribution to knowledge. This is a requirement for a PhD, and many other research projects also make original contributions to knowledge. A vast number of research projects synthesise and analyse existing theory, ideas, and findings of other research, in seeking to answer a particular question or to provide new insights. Such research is often referred to as scholarship; scholarship forms a vital underpinning for almost every type of research project.

Despite its image, research is not an activity which is limited to academics, scientists etc.; it is carried out by everyone many times each day. Some research projects are larger, need more resources and are more important than others.

> **Example**
> Consider what you would do in response to being asked, 'What is the time, please?'
>
> Having understood the question, your response process might be:
>
> - look at watch/clock
> - read time
> - formulate answer
> - state answer ('The time is . . . ').
>
> In providing an answer to the original question, a certain amount of research has been done.

A learning process

Research is a learning process . . . perhaps the only learning process.

Commonly, teaching is believed to be the passing on of knowledge, via instructions given by the teacher, to the learner. Learning is the process of acquiring knowledge and understanding. Thus, teaching exists only through the presence of learning and constitutes a communication process to stimulate learning; teaching is 'facilitation of learning'. If someone is determined not to learn, they cannot be forced to do so, although they may be persuaded to learn through forceful means.

Contextual factors affecting research

Research does not occur in a vacuum. Research projects take place in contexts – of the researcher's interests, expertise and experiences; of human contacts; of the physical environment etc. Thus, despite the best intentions and vigorous precautions, it seems inevitable that circumstances, purpose etc., will impact on the work and its results. The fact that research is being carried out will itself influence the results, as described in the Hawthorne Investigations of Elton Mayo (1949) and noted in the writings of Karl Popper (1989) on the philosophy of research. Research is never a completely closed system.

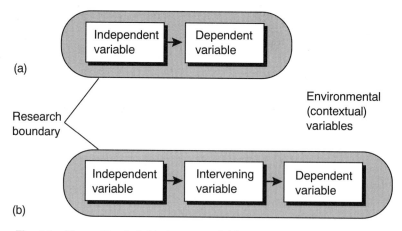

Fig. 1.1 'Causality chain' between variables.

As research is *always* executed in context, it is important to consider the contextual factors, the *environmental variables*, which may influence the results through their impacting on the data recorded. Such environmental variables merit consideration in tandem with the *subject variables* – dependent, independent and intervening (see Fig. 1.1) – of the topic of study. The choice of methodology/methodologies is important in assisting identification of all relevant variables, their mechanisms and amounts of impact.

Example
Consider 'Boyle's Law'. Boyle's law states that, at a constant temperature, the volume of a given quantity of a gas is inversely proportional to the pressure upon the gas, i.e.

$$V \propto \frac{1}{P}$$

$$PV = \text{constant}$$

Laboratory experiments to examine Boyle's Law attempt to measure the volumes of a particular quantity of gas at different pressures of the gas. The temperature is the environmental variable, to be held constant, the pressure is the independent variable and the volume is the dependent variable (following the statement of Boyle's law). The researcher's breathing on the equipment which contains the gas may alter the temperature (otherwise constant) slightly and it will influence the results, though possibly not enough to be recorded.

Boyle's law, like the other gas laws, strictly applies only to a perfect gas, but for many 'practical' purposes, all gases conform to Boyle's law. For this reason, the purpose of the research is likely to be an important determinant of how the experiment is performed and to what level of accuracy. Considerations, such as those noted in respect of Boyle's law experiments, lead to research being classified as pure research and applied research. Slightly different views classify studies as either research or development whilst the purpose of a study often leads to academics' work being classified as research or consultancy.

Classifications of research

Pure and applied research

Frequently, classification of work is difficult, not only due to the use of 'fuzzy' definitions but, more importantly, because the work occurs within a continuum. At one end there is 'pure' or 'blue sky' research such as the discovery of theories, laws of nature etc., whilst at the other, applied research is directed to end uses and practical applications. Most academics are encouraged to undertake research towards the 'pure' end of the spectrum whilst practitioners/industrialists tend to pursue development work and applications.

Essentially, development and applications cannot exist without the basic, pure research whilst pure research is unlikely to be of great benefit to society without development and applications. Unfortunately, much snobbery exists within the research and development sectors – those who work in one sector all too often decry (or fail to value) the contributions of others who work in different sectors. Fortunately, the advances of Japanese industry and many individual organisations which recognise and value the synergetic contributions of the various sectors of the research spectrum are fostering a change in attitude such that research and development activities are recognised as being different and complementary – each with particular strengths, approaches and contributions to make.

Quantitative and qualitative research

The other primary classification system concerns the research methods adopted – broadly, quantitative and qualitative research.

Quantitative approaches adopt 'scientific method' in which initial study of theory and literature yields precise aims and objectives with hypotheses to be tested – conjecture and refutation may be adopted, as discussed by authors such as Popper (1989). In qualitative research, an exploration of the subject is undertaken without prior formulations – the object is to gain understanding and collect information and data such that theories will emerge. Thus, qualitative research is a precursor to quantitative research. In an 'advanced' body of knowledge, where many theories have been developed and laws have been established, quantitative studies of their applicabilities can be undertaken without the need to determine theories and such-like afresh, thereby avoiding, 'reinventing the wheel' for each new study.

Irrespective of the nature of the study, rigour and objectivity are paramount throughout. Commonly, qualitative data, which are subjective data (such as obtained in opinion surveys), can and should be analysed objectively, often using quantitative techniques. However, one should not lose sight of the richness which qualitative data can provide and, often, quantitative data cannot. Triangulation – the use

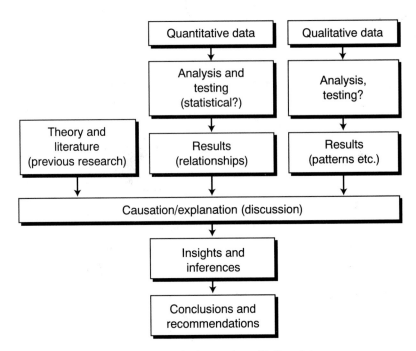

Fig. 1.2 Triangulation of quantitative and qualitative data.

of qualitative and quantitative techniques together to study the topic – can be very powerful to gain insights and results, to assist in making inferences and in drawing conclusions, as illustrated in Fig. 1.2.

Research requires a systematic approach by the researcher, irrespective of what is investigated and the methods adopted. Careful and thorough planning are essential and, especially where large amounts of data are collected, rigorous record keeping is vital – in the study of theory and previous work (literature) as well as in the field work.

The impact of the researcher must be considered, both as an observer, experimenter, etc., whose presence may impact on the data collected and the results derived, and also through bias which may be introduced in data collection, analyses and inferences. Such biases may be introduced knowingly – to examine the subject from a particular viewpoint – or unknowingly, perhaps by asking 'leading questions'.

Example
Consider the question, 'Do you not agree that universities are under-funded?'

The phrasing, 'Do you not agree that…', suggests that the respondent ought to agree that universities are under-funded and so, asking such a question is likely to yield more responses of agreement than if the questions were phrased more objectively.

The question could be phrased much more objectively, 'Do you believe that universities are:

(1) funded generously, or
(2) funded adequately, or
(3) funded inadequately?'

Theories and paradigms

Losee (1993, p. 6) depicts Aristotle's inductive–deductive method for the development of knowledge as shown in Fig. 1.3. He notes that,

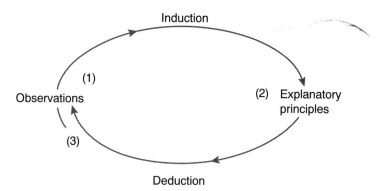

Fig. 1.3 Aristotle's inductive–deductive method (source: Losee 1993).

'scientific explanation thus is a transition from knowledge of a fact [point (1) in the diagram] to knowledge of the reasons for the fact [point (3)]'.

Development of knowledge

Popper (1972, 1989) argues that scientific knowledge is different from other types of knowledge because it is falsifiable rather than verifiable, tests can only corroborate or falsify a theory, the theory can never be proved to be true. No matter how many tests have yielded results which support or corroborate a theory, results of a single test are sufficient (provided the test is valid) to falsify the theory – to demonstrate that it is not always true.

Different philosophies consider that scientific theories arise in diverse ways. Cartesians, who hold a 'rationalist' or 'intellectual' view, believe that people can develop explanatory theories of science purely through reasoning, without reference or recourse to the observations yielded by experience or experimentation. Empiricists, maintain that such pure reasoning is inadequate, it is essential to use experience from observation and experimentation to determine the validity or falsity of a scientific theory. Kant (1934) noted that the scope of peoples' knowledge is limited to the area of their possible experience; speculative reason beyond that, such as attempts to construct a metaphysical system through reasoning alone, has no justification.

Dialectic, a development of 'trial and error', can be traced back to Plato who employed the method of developing theories to explain natural phenomena and followed this by a critical discussion and questioning of those theories; notably whether the theories could account for the empirical observations adequately. Thus, commonly: scientists offer theories as tentative solutions to problems; the theory is criticised from a variety of perspectives; testing the theory occurs, by subjecting vulnerable or criticised aspects of the theory to the most severe tests possible. The dialectic approach, following Hegel and discussed by authors such as Rosen (1982), is that a theory develops through the dialectic triad – thesis, antithesis and synthesis. The theory advanced initially is the thesis; often, it will provoke opposition and will contain weak points which will become the focus of opposition to it. Next, the opponents will produce their own counter-theory, the antithesis. Debate will continue until recognition of the strengths and weaknesses of the thesis and antithesis are acknowledged and the strengths of each are conjoined into a new theory, the synthesis. This is likely to regenerate the cycle of dialectic triad.

History, of course, has a role to play as it is likely to be influential, especially qualitatively, on how people think and behave in developing, criticising and interpreting theories. Popper (1989) uses the term 'historicism', whilst Clegg (1992) considers 'indexicality' to consider history's impact on how people understand, interpret and behave. Indexicality is a person's understanding etc. of terms and is determined by that person's background, socialisation, education, training etc. Marx's broad view was that the development of ideas cannot be understood fully without consideration of the historical context, notably the conditions and situations of their originator(s). It is possible to explain certain social institutions, such as the UK parliament, the Sorbonne, the Supreme Court of USA, or the Tokyo stock exchange, by examining how people have developed them over the years.

Testing a theory

A *theory* is a system of ideas for explaining something; the exposition of the principles of science. Popper (1972) notes four approaches to testing a theory:

- 'The logical comparison of the conclusions among themselves, by which the internal consistency of the system is tested.
- The investigation of the logical form of the theory, with the object of determining whether it has the character of an empirical or scientific theory.
- The investigation of the logical form of the theory, with the object of determining whether it has the character of an empirical or scientific theory.
- The testing of the theory by way of empirical applications of the conclusions which can be derived from it.'

In particular, science provides rules for how to formulate, test (corroborate/falsify) and use theories.

Boolean logic states that concepts are polar in nature – they are either true or false. However, scientific theories are not of that form; they are not well defined, and so it is appropriate to consider a theory as being accepted due to the weight of supporting evidence (until falsified). The value or usefulness of a theory may not be demonstrated by the use of probability alone; such probability must be considered in conjunction with the information contained in the theory. Broadly-based, general theories may be highly probable but vague, due to their low information content; whilst precise or exact theories, with a high information content, may be of much lower probability. Theories with a high information content tend to be much more useful, which leads Blockley (1980) to require that appropriate measures to corroborate theories should be designed such that only theories with a high information content can achieve high levels of corroboration.

Scientific theories must be testable empirically. If a theory is true and one fact is known, another can often be deduced. For example: if a theory states 'all clay is brown' and a sample provided is known to be clay, the deduction is that the sample will be brown. Provided the general statement of the theory is correct, in this case that all clay is brown, the deductive reasoning to go from the general statement to the specific statement, that the sample of clay is brown, is valid. However, discovery of clay which is a colour other than brown will falsify the general theory and so require it to be modified, if not abandoned. Hence, deduction is 'safe', given corroboration of the theory/hypothesis, but it does not allow knowledge to be advanced.

Inductive reasoning – from the specific example to the general statement – is not valid. A *hypothesis* is a supposition/proposition made, as a starting point for further investigation, from known facts. Induction is useful to yield hypotheses, such as that by inspecting a variety of samples it may be hypothesised that all clay is brown. Thus, whilst the hypothesis remain corroborated rather than falsified, deductions can be made from it. Advances are made by use of induction. As knowledge advances, hypothesis may require qualifying statements to be appended to them – such as that all clay of a certain type and found in a given location, is brown – such auxiliary statements lend precision by raising the information content of the hypothesis or theory.

A paradigm

A *paradigm* is a theoretical framework which includes a system by which people view events. The importance of paradigms is that they operate to determine not only what views are adopted, but also the approach to questioning and discovery. Hence, much work concerns verification of what is expected or/and explanation of unexpected results to accord with the adopted, current paradigms. As progressive investigations produce increasing numbers and types of results which cannot be explained by the existing paradigms' theoretical frameworks, paradigms are modified or, in more extreme instances, discarded and new ones adopted – the well-known 'paradigm shift'.

Normally, the advance of knowledge occurs by a succession of increments, hence it is described as evolutionary. Only rarely are discoveries made which are so major that a revolutionary advance occurs. Often, such revolutionary advances require a long time to be recognised and more time, still, for their adoption, such as Darwin's theory of evolution. Hence, in terms of scientific progress, a theory which is valid at a given time is one which has not been falsified, or one where the falsification has not been accepted. Whilst objectivity is sought, research does have both cultural and moral contents and so a contextual perspective, especially for social science research, is important to appreciate the validity of the study.

Models and hypotheses

A primary use of theory is to facilitate prediction. Instances where theories fail to predict correctly are anomalies. However, if a number of serious anomalies occur, the theory is likely to be rejected in favour of one which is more appropriate: one which explains all the occurrences more accurately. Hypotheses lead to theories which may be modified by auxiliary statements which 'particularise' them. They may be eventually rejected in favour of another theory of wider, accurate predictive ability. During the period of modifications and potential substitutions of theories, the 'competing' theories may be the subject of much debate in which advantages and disadvantages of each are considered to yield hierarchies of the theories continuously.

Another great value of theories is to enable researchers to produce models which show how the variables of a theory are hypothesised to interact in a particular situation. Such modelling is very useful in clarifying research ideas and limitations and to give insights into what should be investigated and tested.

Research styles

Bell (1993) suggests styles of research to be Action, Ethnographic, Surveys, Case Study and Experimental. Unfortunately, definitions of such styles vary and so, at best, the boundaries between the styles are not well defined.

Action research

Generally, action research involves active participation by the researcher in the process under study, in order to identify, promote and evaluate problems and potential solutions. Inasmuch as action research is designed to suggest and test solutions to particular problems, it falls within the applied research category, whilst the process of detecting the problems and alternative courses of action may lie within the category of basic research. The consideration of quantitative v. qualitative categories may be equally useful.

Ethnographic research

The ethnographic approach demands less active 'intrusion' by the researcher and has its roots in anthropology. The researcher becomes part of the group under study and observes subjects' behaviours, statements etc. to gain insights into what, how and why their patterns of behaviour occur. Determination of cultural factors such as value structures and beliefs may result, but the degree of influence caused by the presence of the researcher, and the existence of the research project, will be extremely difficult (if not impossible) to determine.

Complementarily, a sociological or political perspective recognises that the investigator is part of the group being studied, and so is a viable member of the group and a participant in the group behaviour as well as being the observer – more akin to the action research approach.

Surveys

Surveys operate on the basis of statistical sampling; only extremely rarely are full population surveys possible, practical or desirable. The principles of statistical sampling – to secure a representative sample – are employed for economy and speed. Commonly, samples are surveyed through questionnaires or interviews. Surveys vary from highly structured questionnaires to unstructured interviews. Irrespective of the form adopted, the subject matter of the study must be introduced to the respondents. For a given sample size of responses required, particular consideration must be given to the response rate (i.e. the percentage of subjects who respond) and number of responses obtained. Following determination of the sample size required, appropriate procedures must be followed to assist in securing the matching of responses to the sample selected. This is a special consideration for 'stratified' samples; samples classified into categories, usually by size or measured degrees of some important, continuous attribute.

Case studies

Case studies encourage in-depth investigation of particular instances within the research subject. The nature of the in-depth data collection may limit the number of studies, when research is subject to resource constraints. Case studies may be selected on the basis of their being representative with similar conditions to those used in statistical sampling to achieve a representative sample, to demonstrate particular facets of the topic, or to show the spectrum of alternatives. Case study research may combine a variety of data collection methods, with the vehicle or medium of study being the particular case, manifestation or instance of the subject – such as a claim, a project, a batch of concrete.

Commonly, case studies employ interviews of key 'actors' in the subject of study; such interview data may be coupled with documentary data (such as in a study of a production process). Alternatively, a case study may be 'situational', such as a wage negotiation or determining safety policy, and for such research, several 'cases' may be studied by individual or combined methods of ethnography, action research, interviews, scrutiny of documentation etc. Hence, although case studies constitute a distinct 'style' of research, they are a means of obtaining data rather than a particular methodological approach.

Experiments

The experimental style of research is, perhaps, suited best to 'bounded' problems or issues in which the variables involved are known, or at least hypothesised with some confidence. The main stages in experimental design are shown in Fig. 1.4. Usually, experiments are carried out in laboratories to test relationships between identified variables; ideally, by holding all except one of the variables constant and examining the effect on the dependent variable of changing the one independent variable. Examples include testing the validity of Boyle's Law, Hooke's Law, and causes of rust experiments. However, in many cases, notably in social sciences and related subject fields, experiments are not conducted in specially built laboratories but in a dynamic social, industrial, economic, political arena. An example is Elton Mayo's 'Hawthorne Experiments' which took place in a 'live' electrical manufacturing company (Mayo, 1949).

Fig. 1.4   Experimental design.

Example

Consider investigating client satisfaction with the provision of a construction project. What quantitative and what qualitative data are likely to be available readily on a case study of a construction project?

Quantitative data would consider time and cost performance derived from project records – predicted v. actual; quality might be considered from records of re-worked items, corrections required due to defects recorded during the maintenance period – measured by number, value etc.

Qualitative data could present participants' perceptions of client satisfaction in respect of the performance criteria of cost, time and quality. Such data would be obtained through questioning of those participants' identification of the variables and hypothesising of their inter-relations. Research can proceed by holding all but one of the independent variables constant and examining the effects of controlled changes in the remaining independent variable on the dependent variable.

In certain contexts, such as medical research, the sample under study may be divided into an experimental group and a control group. After the experimental period, the groups may be compared, to determine any differences between the groups which can be attributed to the experiment. In such cases, the members of the groups must not know to which group they belong; it is helpful also (to avoid possible bias in analysis), if those who carry out the analysis of results are not informed of which person is in each group either.

Hence, experimentation is aimed at facilitating conclusions between cause and effect – the presence, extent etc. Experimentation is at the base of scientific, quantitative method.

Quantitative and qualitative approaches

It is quite common for small research projects to be carried out with insufficient regard to the array of approaches which may be adopted. This may be because the appropriate approach is obvious, or that resource constraints preclude evaluation of all viable alternatives or it may be due to a lack of awareness of the alternatives. Such lack of awareness does not mean that the research cannot be executed well, but often it does mean that the work could have been done more easily and/or could have achieved more.

Research methods and styles are not usually mutually exclusive although only one, or a small number of approaches, will normally be adopted due to resource constraints on the work. The different approaches focus on collection of data rather than examination of theory and literature. The methods of collecting data impact upon the analyses which may be executed and, hence, the results, conclusions, values and validity of the study.

Quantitative approaches

Quantitative approaches seek to gather factual data and to study relationships between facts and how such facts and relationships accord with theories and the findings of any research executed previously (literature). Scientific techniques are used to obtain measurements – quantified data. Analyses of the data yield quantified results and conclusions derived from evaluation of the results in the light of the theory and literature.

Qualitative approaches

Qualitative approaches seek to gain insights and to understand people's perceptions of 'the world' – whether as individuals or groups. In qualitative research, the beliefs, understandings, opinions, views etc. of people are investigated – the data gathered may be unstructured, at least in their 'raw' form, but will tend to be detailed, and hence 'rich' in content and scope. Consequently, the objectivity of qualitative data often is questioned, especially by people with a background in the scientific, quantitative tradition. Analyses of such data tend to be considerably more difficult than with quantitative data, often requiring a lot of filtering, sorting and other 'manipulations' to make them suitable for analytic techniques.

Analytic techniques for qualitative data may be highly laborious, involving transcribing interviews etc. and analysing the content of conversations. Clearly, a variety of external, environmental variables are likely to impact on the data and results and the researchers are likely to be intimately involved in all stages of the work in a more active way than usually is acceptable in quantitative studies.

Triangulated studies

Both qualitative and quantitative approaches may adopt common research styles – it is the nature and objectives of the work together with the (consequent) nature of the data collected which determine whether the study may be classified as qualitative or quantitative. Given the opportunity, of course, triangulated studies may be undertaken. As triangulated studies employ two or more research techniques, qualitative and quantitative approaches may be employed to reduce or eliminate disadvantages of each individual approach whilst gaining the advantages of each, and of the combination – a multi-dimensional view of the subject, gained through synergy.

Whatever approach, style or category of research is adopted, it is important that the validity and applicability of results and conclusions are appreciated and understood. In particular, it is useful to be demonstrably aware of the limitations of the research and of the results and conclusions drawn from it. Such limitations etc. are occasioned by various facets of the work – sampling, methods of collecting data, techniques of analysis – as well as the, perhaps more obvious, restrictions of time, money and other constraints imposed by the resources available. Hence, it is very helpful to consider the constraints, methods etc. at an early stage in the work to ensure that the best use is made of what is available. Indeed, it may well be preferable to carry out a reduced scope study thoroughly than a larger study superficially – both approaches have validity but achieve different things.

Data sources

As with any project, the planning phase is crucial and it is wise to evaluate what is being sought and the alternative approaches available as early as possible. Of course, re-evaluations may be necessary during the course of the work, in instances such as where the data required prove to be unavailable. As data are essential to research, it is useful to consider what data are required, and alternative sources and mechanisms for collection during the planning phase. Use of surrogate data (indirect measures of what is sought) may have to be used, especially where the topic of study is a sensitive one (e.g. cost, safety, pricing, corruption, labour relations).

Where researchers have good contacts with potential providers of data, use of those sources is likely to ease the data collection process. If trust and confidence have been established, it is likely to be easier to obtain data and it may be possible to obtain data which might not be available otherwise. Certainly, trust and confidence are important considerations in data collection – the more sensitive the data, the more trust in the researcher which is required by the provider. Often, it is essential to ensure that the providers of data cannot be traced from the output of the research. Statements ensuring anonymity are helpful as are methods which demonstrate anonymity in the data collection methods, such as not requiring names and addresses of respondents. However, anonymity must work. It is hardly providing anonymity if one identifies respondents as A, B ... N rigorously in the research report but thanks respondents by name in the acknowledgements section.

Occasionally, respondents wish to scrutinise a report prior to its 'publication'. Whilst such provision is useful in building confidence over anonymity issues, and may assist in ensuring accuracy of data etc., it may be regarded as an opportunity for the respondents to comment on the research, and possibly to demand changes – perhaps, to remove portions with which they disagree or which they dislike. Such changes should be resisted, provided the research has been conducted properly, as they would distort the research report and, thereby, devalue the work.

For applied research, it is increasingly popular to form a *steering group* of the principal investigators, industrialists and practitioners. The steering group helps to form the strategy for the execution of the work and to monitor and guide the research during its execution. The objective is to ensure the combination of rigorous research with practical relevance. Of course, there are spin-off benefits of the researcher's enjoying easier access to data via the committed practitioners, and the practitioners' gaining knowledge and insight of issues/problems which are important to them.

Where to begin

Research methodology refers to the principles and procedures of logical thought processes which are applied to a scientific investigation.

Method concerns the techniques which are available and those which are actually employed in a research project. Any management of a research project must address certain questions in making decisions over its execution. The questions involved are:

- what?
- why?
- where?
- when?
- how?
- whom?
- how much?

It is those questions which study of this book will assist in answering or, rather, provide some information to help to reach an answer. By addressing the issues explicitly and logically, noting requirements, constraints, and *assumptions*, the progress through research projects will smooth and ease progressively as expertise and experience develop.

Often, a researcher is able to select a supervisor or supervisors. In selecting a supervisor, three considerations apply – that person's subject experience and expertise, research experience and expertise and, perhaps the most important factor differentiating potential supervisors, the ability to relate to and communicate well with the researcher. The best research tends to be executed by people who get on well together as well as possessing complementary skills and expertise.

It is important to determine the scope of the work at the outset; the most common problem is for a researcher to greatly over-estimate what is required of the work, what can be achieved and the amount of work that can be done. It is a good idea to consult an experienced supervisor or 'third party' to ensure that the programme and scope of the research intended is *realistic*.

Example
What? concerns selection of the topic to be researched with consideration of the level of detail. It is useful to note the resources available and constraints so that an appropriate scope of study can be determined.

Why? may command a variety of answers, each of which applies individually but some of which may apply in combination. So, 'required for a degree', 'required by employer', 'interest', 'career development', and possibly many other reasons, may be advanced to say why research is being undertaken. However, why a particular research project is being carried out or proposed, apart from the reasons given already, may be due to its being topical or because the researcher has expertise in that subject and wishes to use that expertise to acquire and advance knowledge in that field.

Where? Obviously all research occurs somewhere – the host institution may be a university, as well as the various places at which individual research activities occur – libraries, data collection points, visits to experts etc. It may be useful to consider the amount of travel, both cost and time, as an input to the strategy for executing the research.

When? The timing of the research and time available to carry it out will usually be specified. It will be necessary to produce a timetable for the work by dividing the time available between the component activities. Often there will be restrictions on the time for data collection – allow for holiday periods, very busy periods etc.; what sequences of activities are necessary and what are the alternatives? To what extent can the activities overlap? A common problem is to devote insufficient time to planning the work and to forget, or at least to under-estimate, the time necessary for data analyses, production of results and conclusions and for preparation of the report. All too often the only real focus is on fieldwork (data collection) – such enthusiasm is healthy but must be kept under control.

How? – is the issue of methodology. In some instances, the methodology is obvious – virtually 'given' – as in computational fluid dynamics. Commonly, a topic may be investigated in a variety of ways, individually or in combinations, so a choice must be made. The choice will be influenced by the purpose of the research, the subject paradigm, the expertise and experience of the researcher and supervisor (if any), as well as practical considerations of resource and data availabilities.

Whom? Four main groups of people are involved in the execution of research – the researcher, the supervisor, the sample personnel, who provide the data or access to it, and others who can help, such as laboratory technical staff. Naturally, a research project is 'commissioned' by someone, for instance, a university as a requirement of a course of study, an academic agency such as a research council or a commercial agency, perhaps a government body, company consultant practice etc.

How Much? This issue concerns the resources which can be used. Many resources, such as money, are fixed but people's time tends to be rather flexible – especially the time input by researchers themselves. No research project is really completed from the researcher's point of view as there is always a bit more which could or ought to be done. Hence each report contains a list of recommendations for further research.

Summary

This chapter has introduced some main concepts of research. A definition of research has been provided and the contexts of undertaking research have been discussed. Various approaches to research have been examined – notably qualitative and quantitative – together with their combination through 'triangulation'. Styles of study have been considered and questions which research projects address have been noted. The issues of confidentiality and anonymity have been discussed and the essential need for objectivity has been emphasised.

References

Bell, J. (1993) *Doing your Research Project*, 2nd edn, Open University Press, Buckingham.

Blockley, D.J. (1980) *The Nature of Structural Design and Safety*, Ellis Horwood Ltd, Chichester.

Clegg, S.R. (1992) Contracts cause conflicts. In: *Construction Conflict Management and Resolution* (eds P. Fenn & R. Gameson), pp. 128–144, E. & F.N. Spon, London.

Kant, I. (1934) *Critique of Pure Reason* (trans. F.M. Muller), 2nd edn, Macmillan, New York.

Losee, J. (1993) *A Historical Introduction to the Philosophy of Science*, 3rd edn, Opus, Oxford.

Mayo, E. (1949) *The Social Problems of an Industrial Civilisation*, Routledge & Kegan Paul, London.

Popper, K.R. (1989) *Conjectures and Refutations: The Growth of Scientific Knowledge*, Routledge & Kegan Paul, London.

Popper, K.R. (1972) *The Logic of Scientific Discovery*, Hutchinson, Oxford.

Rosen, M. (1982) *Hegel's Dialectic and its Criticism*, Cambridge University Press, Cambridge.

Chapter 2

Topic for Study

> The objectives of this chapter are to examine the process of:
>
> - **selecting a topic**;
> - **writing a research proposal.**

Selection of a topic

Very often, the most difficult task for any researcher is to select a topic for study and then to refine that topic to produce a proposal which is viable. Generally, people set targets far too high in terms of both the extent of the research which is possible and the discoveries which are sought. It is surprising to most new researchers how little (in scope) can be achieved by a research project, and hence the necessity to restrict the study in order that adequate depth and rigour of investigation of the topic can be undertaken.

Resources

An important aspect to evaluate is the quantities of resources which can be devoted to the study. Often it is helpful to calculate the number of person-hours, days, weeks, months or years, which are available for the research. Given a fixed amount of time and the period within which the research must be completed, and taking account of any flexibilities, the amount of work which can be undertaken begins to be apparent. Usually a report of the work is required, and that report must be produced within the time frame, so the period required to produce the report reduces the time available for executing the study itself.

Many people consider that undertaking a research project is 2% inspiration and 98% perspiration – clearly, research is *not an easy option*. Research is hard work, but it is often the most rewarding form of study. The satisfaction and sense of achievement derived from a project completed well can be enormous; the efforts are well worthwhile and provide the researchers with expertise and insights for future work. However, especially in the early days of a project, enthusiasm is a great asset – it is a major contributor to overcoming difficulties which will almost inevitably arise. Determination is valuable for a researcher as it will help to ensure that the project is seen through to completion.

Even in cases where a topic is given – such as where a researcher applies for a post to carry out a particular project advertised – there is selection of the topic by the prospective researcher. Generally, where a research project is part of a course of study, the choice of topic to research is made by the individual *but* that choice should not be made in isolation. Potential tutors, supervisors and mentors should be consulted, together with colleagues and, if possible, practitioners to assist in selecting a topic which is interesting, viable and appropriate to the context and the people concerned – most especially the *researcher*. The amount of time and effort spent in selecting and refining a topic and then planning to yield a proposal may appear very long, if not excessive. Invariably, it will be time well spent. Such formative stages are of paramount importance, and often will be the main factor determining whether the research is a success.

Some academics believe that the Pareto distribution applies to research study. A Pareto distribution is the '80-20 rule' (Fig. 2.1); a small proportion of components have the major effect on the outcome. Applied to a production project, about 20% by number of the components account for about 80% of the project cost. The Pareto distribution is believed to apply far and wide; it applies to

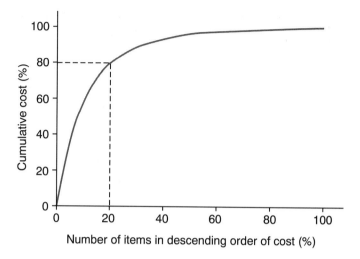

Fig. 2.1   Pareto distribution applied to research.

programmes of study in that 80% of the work is completed (or becomes visible) in the last 20% of the time available – this is partly due to preparatory work being carried out in the early part of the study and so not being accessible, but also because certain people do not do the work until the deadline looms, due to other pressures or lack of programming. Requesting an extension of time may not be viewed favourably; indeed, if such a Pareto distribution does apply, it may be preferable not to grant an extension.

Subject selection

In most cases, researchers must confront the issue of subject selection. Although the selection will not be made from all subject areas but will be confined within the boundaries of particular disciplines, the possibilities are vast. Therefore, it is helpful to consider the process of subject selection as one of progressive narrowing and refinement – the degree to which those are taken being determined by the nature of the discipline and the appropriate research methods. Essentially, the process is strategic – the particular initial issue is *what*?

In deciding any course of action, it is a good idea to undertake a SWOT analysis in which PEST factors are scrutinised. SWOT analysis requires determination of Strengths, Weaknesses, Opportunities and Threats. Strengths and Weaknesses are features of the individual or organisation and are *internal* factors, whilst Opportunities and Threats are present in the environment, so they are external, or exogenous, factors. PEST factors are the Political (and legal), Economic, Social and Technical forces in the environment. Consideration of the PEST factors assists analysis of the Opportunities and Threats. For examples see Fellows *et al.* (1983) and Newcombe *et al.* (1990). A sound strategy is to build on strengths and overcome weaknesses in seeking to take advantage of opportunities and to minimise the possible effects of threats.

In selecting a subject for a research project, it is useful to begin by constructing certain lists:

List 1: Topics of interest.
List 2: Personal strengths and weaknesses.
List 3: Topics of current interest in practice.
List 4: Access to data.
List 5: Research limitations.

The first list may contain topics of interest in quite broad terms. A second list might be of personal strengths and weaknesses. Certain strengths and weaknesses could influence the choice of topic directly whilst others might do so by consideration of the methodology and the nature of the analyses which would be appropriate. However, whilst building on strengths is good, an individual, interest-driven research project may be an excellent vehicle to extend the researcher's knowledge and experience. So, there are good arguments for pursuing a topic of research which is of particular interest to the researcher – it could serve to enhance career opportunities.

A third list might concern topics of current interest – what is being debated in the technical press – the 'sexy' or 'hot' topics of the day. Adopting such a topic for study should ease data collection owing to the amount of interest in the subject. However, interest in topics tends to be quite short-lived and the clever thing to do is to predict what the next issue of debate will be. In the UK construction industry, hot topics seem to have a 'life' of less than two years, although many 'cycle round' again and again.

Example

List 1 Begin by listing topics of interest in quite broad terms e.g. Concrete prestressing; Project procurement approaches; Price setting mechanisms. Such topics are too broad and insufficiently specific for certain types of research, but topics such as 'The culture of design and build contractors' or 'Effect of water content on bearing capacity of London Clay' could lend themselves to investigation.

List 2 A second list on strength and weaknesses may contain self-evaluations such as 'good at economics but quite weak in maths and statistics' or 'experienced in new build and in contracting but no experience in refurbishment or design consultancy'.

List 3 Over recent years, 'hot topics' have included; buildability, sick building syndrome, life cycle costing, risk management, value management, quality assurance, quality management, refurbishment, facilities management, dispute resolution procedures. Other topics, although never getting quite 'red hot', may retain enduring interest; CAD, communications, information flows, payments mechanisms, inter-party relationships, price determination, environmental issues, client satisfaction, demolition of prestressed tower blocks.

Further lists may concern access to data via industry or practice contacts. A particular employer may wish for issues to be investigated. Such projects should facilitate co-operation in data provision but care is needed about restrictions on publication of the contents of the report etc. Other lists may include the interests and expertise of potential supervisors and collaborators, and other researchers' work to which access can be gained readily.

Finally, it is likely to be helpful to list constraints and the resources available – these lists will be invaluable in determining the practicality of a proposed study and assist in determining what can be done, rather than trust what it would be 'nice' to do.

Example

List 4 Access to data should ideally match the data required, but is likely to be modified by practicalities of obtaining the data; especially if the data are 'sensitive' (e.g. costs, safety,

> corruption). Employers and sponsoring bodies as well as
> professional institutions may be helpful in securing access to
> the data needed.
>
> **List 5**   Research limitations will be: the constraints on the resources
> available to execute the research and, hence, the scope of
> the study; data available; the methods/techniques employed.
> The time and resources available for the research, especially
> for dissertations and academic theses, are likely to be well-
> known, so it is the particular limitations which relate to the
> research which should be noted – especially data availabil-
> ity. However, if limitations should have been foreseen from
> the nature of the topic – sensitivity, the literature, experi-
> ences of other researchers etc. – care must be taken to
> explain why such limitations were not taken into account and
> avoided in the research design. In any event, it is important to
> note the reasons for what was done to overcome the
> limitations and to ensure validity of the resultant research.

It is best if the lists contain 'raw' ideas and do not contain the results
of sophisticated evaluations – they come later. Essentially, the lists
result from 'brainstorming' so it may be helpful to do the brainstorming in a group designed in such a way that people's ideas
'spark' further ideas in others – no evaluations or disparaging com-
ments, just throw out the ideas, note them and think about further
possible topics until no more emerge. Do not force things; if it is
difficult to get a first idea, it may be helpful to visit the library, take
a book at random, open it and pick a word at random and see if a
research topic can be devised from that word. Such a serendipity
approach may yield a novel and exciting topic (if horse races were
predictable, everyone should win but would that be exciting?).

Choosing a topic

Having produced a number of lists of possible topics, constraints,
requirements etc., the next step is to begin some evaluation. Again,

it is important to let the research topics be the driving force and the requirements and constraints be the parametric factors, denoting the limits and extent of what may be done – they should not be the dictating factors even though they may 'loom large' from a student researcher's perspective.

Evaluating alternatives

Within overall subject domains, a number of topics will have emerged. As a background to the evaluation process, it may be helpful to consider two issues:

- 'What does the research seek to achieve?'
- 'What does the research seek to find out?'

In terms of achievement, apart from the obvious, important, but also mundane issue of satisfying the imposed purpose of carrying out the project, such as obtaining a good grade to contribute to a degree, the personal advancement aspects are important. Broadly, the achievement may be to extend understanding in a particular subject area, to broaden understanding in a new subject area by enhancing the depth of knowledge of the researcher or to gain knowledge in subjects where the researcher has less expertise. Of course, combinations of achievement in both ways may be possible, but the underlying issue is largely one of self-motivation, to cope with the 98% perspiration (hard work).

The second issue – what the research sets out to find out – concerns selecting a particular topic within the overall subject area decided already. So, the list of topics allocated to the overall subject area selected will be evaluated. Again, the lists of personal factors, data availability, etc. will be used in the evaluation, but retaining the approach of the topic as driver and the various constraints as parameters. It may be difficult to make a single choice, but whether one, two or three topics are the result of the evaluation, it will be helpful to consult literature – leading journals and reputable texts – to gain some preliminary, but further, understanding of the issues relating to the topics, whether they have been investigated extensively already and any likely problems. Secondly, it is useful to examine what particular terms are used, what the variables are, what data are needed and how those data may be collected.

A researcher should not be dismayed if the topic or closely related ones have been investigated previously. Much research, including scientific method, uses not only conjecture and refutation but, as an integral part of that philosophy, looks to replication of research. Studies of safety or performance in the UK construction industry are worthy of replication in Hong Kong, Malaysia, USA, China etc., partly to evaluate the methodology in other contexts, partly to re-examine the findings of the original study and partly to examine the situation, both absolutely and comparatively, between the locations of the studies. Indeed, replication, even if on a reduced scale, of a leading study, but in a different context, thereby changing only one variable if possible, has the great advantage of using a tested and substantiated methodology. This allows more emphasis to be placed on examination of results, whilst monitoring the methodology and providing any necessary critique due to the changed circumstances.

So, evaluations of the topics from the perspectives of desires and constraints, often in discussion with colleagues, practitioners and the potential supervisor, will result in the selection of a single topic. The result of such an evaluation does not mean that details of the topic will not change. Research, by its nature, is a 'voyage of discovery', an iterative learning process, and investigation of the unknown is likely to require changes from *a priori* expectations. If it were not so, it might not be research, although the extent of changes should not lead to radical changes in the topic.

Refining a topic

The process of selection will continue for some time as investigations proceed and the topic (through considerations of definitions, variables and their relationships, aspects of theory, findings of previous work etc.) emerges and undergoes progressive refinement. The goal is to reach a state in which the topic is delineated sufficiently well for the aim and objectives of the research to be identified, and appropriate methodologies considered, to enable a research plan to be formulated, including a draft timetable.

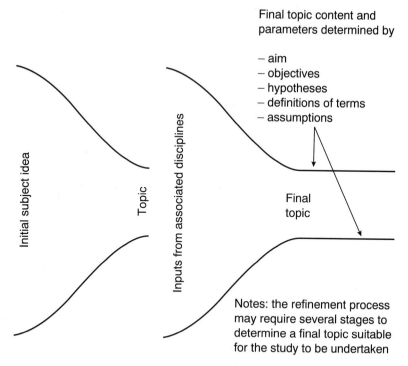

Final topic content and
parameters determined by

− aim
− objectives
− hypotheses
− definitions of terms
− assumptions

Final
topic

Initial subject idea

Topic

Inputs from associated disciplines

Notes: the refinement process
may require several stages to
determine a final topic suitable
for the study to be undertaken

Fig. 2.2   The process of refining a topic for research.

The process of refining a topic for research is depicted in Fig. 2.2.

Writing the proposal

The outcome of the initial considerations and investigations will be a
proposal for the research. The UK's Engineering and Physical
Sciences Research Council (EPSRC) prescribes the format for
proposals, as do most funding agencies. It comprises a form concerning the support requested and an outline of the project plus a six page
'case for support' outlining the project proposal in more detail.

Normally, for a degree of Bachelor or Master dissertation, a
proposal of around four sides of A4 is adequate; a proposal for
M.Phil or PhD will be more extensive, but all proposals should be
concise. Depending on the nature of the research proposed, the
proposal should contain:

- Title
- Aim
- Objectives
- Hypothesis
- Methodology
- Programme
- List of primary references.

Usually, it is helpful to append a diagram showing how the variables envisaged are relevant to the research and how they are hypothesised to relate to each other. Often, that forms the basis of very helpful analytic discussions of the topic and aids in-depth thinking and investigation throughout.

Aim

The aim of a research project is a statement of what the research will attempt to do – often in the form of what is to be investigated, which is more appropriate for qualitative research, or what impact the main independent variables are believed to have on the dependent variable, an approach which is more suitable for quantitative studies. The aim is really a statement at the strategic level so, usefully, can be considered to be what the researcher would like to do if resource constraints and other constraints did not exist. Clearly, constraints do exist, and so the research should not be judged, once completed, against the yardstick of the statement of aim. Rather, the aim provides the identification of the context of what is attempted.

> *Example*
> To investigate the 'maintenance path' for local authority school buildings in UK through establishing maintenance needs and work execution mechanisms, and provide maintenance information to designers in an environment of resource constraints.

Objectives

The objectives are statements within the strategic statement of aim; they are statements at the tactical level. Objectives take the aim of the research and, given the constraints, translate the aim into coherent, operational statements. These are statements which relate to each other logically but which are, each, self-sufficient also and describe what the research hopes to achieve or discover through the study.

The objectives specify what it is hoped will be discovered by the research – what will be known at completion of the project which was not known at its start *and* has been revealed by the research. However, for qualitative studies, the objectives may concern how the study will be undertaken and some details of what is to be studied. Both approaches are valid but they are different. It is important for researchers to recognise the differences and the consequences for the research resulting from the adoption of the alternative approaches. So, say, for a research project to investigate the impact of the culture of participants on the performance of construction projects, it may be useful to conduct a dual investigation – a sociologically based study providing a more qualitative approach in parallel with a construction management based study giving a more quantitative approach. Results of the individual investigations could perhaps be triangulated to yield synergy.

For ethnographic-type research, a broad statement of aim may be all that can be produced. It may be that in some instances, even that may not be viable – only identification of the subject area may be possible. The researchers will have notions of what is to be investigated, but due to lack of theory, prior investigation etc. they may not produce detailed objectives and hence, methodology. The approach is to collect all possible data rigorously and to use those data to structure the study from the relationships and patterns which emerge. It is probable that all knowledge began to be discovered from this approach – perhaps most obviously in the social sciences where people's behaviours had to be studied to detect patterns etc. from which hypotheses were generated for testing and the subsequent derivation of theories and 'laws'.

For most research projects, especially smaller ones, it is good discipline to restrict the project to a *single* aim and the objectives to about *three*. Such restriction promotes rigour in considering what the research is about and what can be achieved realistically. Keep the statements simple, especially for objectives, working with one independent variable in each which impacts on the dependent variable of the study, and ensuring that those variables can be identified.

> *Example*
> 1. To investigate any linkages between construction types and maintenance requirements.
> 2. To examine any relationship between age of buildings and their maintenance needs.
> 3. To determine the factors which impact on maintenance work execution for UK local authority school buildings.
> 4. To develop and test a model for maintenance of UK local authority school buildings.

Hypotheses

Certain studies, such as ethnographic research projects, may achieve the formulation of an hypothesis as their end result. Other studies may develop hypotheses as the result of initial investigations, whilst a third category will formulate hypotheses in the early stages of the work – as in experimental research projects. Irrespective of when an hypothesis is formulated, it is a statement of conjecture – it suggests a relationship between an independent and a dependent variable and the nature of that relationship. The statement concerns direction in the relationship, known as causality.

> *Example*
> Consider the following hypothesis:
>
> 'The method of programming construction projects employed by contractors influences project performance, and hence participants' satisfaction with those projects.'
>
> Apart from criticisms of the English and phrasing of the hypothesis, it contains two dependent variables – project performance and participants' satisfaction. This raises issues of what may be said about support for the hypothesis if, after testing, one part is supported and the other is not. Clearly, it would be preferable to split the hypothesis into two, or even three:-
>
> - programming–performance
> - performance–satisfaction
> - programming–satisfaction.

> The performance–satisfaction relationship is implied in the hypo-
> thesis. To retain this in the study, it could be determined from theory
> and previous work but, for rigour and completeness, it should form an
> element of the research.

Hypotheses are statements which are produced to be tested objectively. It is useful to have *one* main hypothesis, derived from the aim of the research, and sub-hypotheses relating to the objectives (if appropriate). Further sub-hypotheses may be included provided that they assist or clarify the research; too many will tend to promote confusion. Where appropriate, hypotheses are extremely valuable in lending direction, constraints of relevance and objectivity to a research project. The goal must be to *test* the hypotheses objectively through data collection and analyses; usually, hypotheses are derived from theory and literature.

A common misconception is that once hypotheses have been established, the goal of the research is to prove them. That is neither possible nor desirable. Even the most extensive research will not prove anything absolutely, although it may establish a likelihood with an extremely high level of confidence, as is common in medical research. Seeking to prove or to disprove an hypothesis is likely to introduce bias into the research – contrary to the requirement of objectivity. So, if a researcher believes the hypothesis to be true, it may be a good idea to propose the opposite hypothesis to attempt to counter any bias in the researcher's initial beliefs, as there may be an innate tendency for researchers to seek to support the hypothesis of the study unwittingly.

Hypotheses also focus the work on relevant aspects and help to identify boundaries of the study in experimental and quantitative researches. Especially where resources are very limited, it is invaluable to be able to identify the boundaries for the study to ensure that effort is expended only where it will be relevant to the particular topic. Such an approach is alien to much qualitative work which seeks immersion in the subject matter to collect all possible data for analysis to see what, if anything, emerges.

Use of hypotheses indicates data requirements and suggests analyses to be performed. Naturally, other relationships may emerge, and so it is important to retain openness to new ideas etc. which emerge. Although tests should examine what has been hypothesised, there must always be a preparedness and opportunity for other findings to emerge – comprehension rather than restriction must apply.

Methodology

Methodology, the methods by which research can be carried out, lies at the heart of research. Many good ideas remain uninvestigated and/or unfunded because the methodology has not been considered adequately. Many of the following sections are concerned with the major aspects of research methodologies and the detailed considerations of these have been left until later. It is vital that the methodology is given careful consideration at the outset of the research so that the most suitable approaches and research methods are adopted. Whenever possible, it is useful to draw a diagram of the variables likely to be involved and the hypothesised relationships between them. In determining and considering the methodology for research, attention should be given to 'DATA'; namely :

D   Definitions of the main terms involved; especially where terms have varied definitions, it is essential to decide *explicitly* the definitions to be adopted, and why they have been adopted, so that appropriate measurements can be made during collection of data.

A   note the Assumptions which are made and the justification for them; if possible, explain their consequences and examine what occurs when any assumptions are relaxed.

T   research, and critically review the Theories, principles and literature relating to the subject matter of the research.

A   evaluate what Analyses may be carried out with respect to data available, the objectives and any hypothesis, so that the most robust and rigorous analytic methods will be used, thereby maximising confidence in the results.

Programme

Milton Friedman, the Chicago-School economist, stated that, 'Only surprises matter'; perhaps that statement is more appropriate to research than to any other activity.

The production of a timetable or programme for any research project is essential to ensure the project's viability. It helps the researcher, and others, to decide how the activities should fit together and the time to be devoted to each one. Thus, once formulated and agreed, it will provide a basic yardstick against which progress may be monitored. However, due to the nature of research, it is important that the programme, and those using it, have sufficient flexibility to enable novel, potentially productive, lines of investigation to be noted or pursued in some way.

Normally, a simple bar chart showing the main research activities is adequate as a programme for the work. For smaller, quantitative projects, times for 'producing the report', 'finalising the proposal', 'theory and literature review' and 'data analysis' may be predicted with some confidence; programming these activities will indicate the amount of time and timing of the data collection. Despite overlaps of activities, the time available for the field work and analysis is likely to be quite constrained, so efficiency and effectiveness of data collection are essential.

Deliverables and industrial or practitioner support

Increasingly, applied research is required to focus on the provision of practically-useful deliverables. On many fronts, research is being pushed, via the funding agencies, towards serving the needs of industry and commerce. Fundamental and qualitative research is unlikely to produce immediate, industrially useful deliverables, such as results and findings, directly although, in the long term, results of such work can be extremely important and far-reaching.

In Britain and many other Western market-developed economies, industry and commerce tend to have short term views, and so will support research projects only if they appear likely to produce profit-enhancing results quickly. This may perhaps be due to performance imperatives imposed by financiers. Such a view is likely to present less problems for short duration, applied research than for long duration, fundamental studies. However, support there must be, if only to facilitate collection of data. In seeking support, it is helpful to show the relevance of the work and to contact a person likely to have interest in the study and to be in sympathy with it. The support of an industrial or professional association will help. The offer of the provision of a summary of findings is not only good manners in showing appreciation of assistance and it can be given confidentiality, if necessary. Do ensure that such summaries *are provided* to the participants. Letters of request from the researcher's institution lend credibility to the work, ensure legitimacy of the study and should provide appropriate and sympathetic 'control' over sensitive facts.

Summary

This chapter has considered the processes by which a suitable topic for research, such as a dissertation, may be selected. Often, the selection process is one of narrowing from a subject area to a particular issue for study. Criteria and parameters must be evaluated along with the rationale for undertaking the research. Throughout, the requirement is for objectivity, especially in the formulation and subsequent testing of hypotheses.

References

Fellows, R.F., Langford, D.A., Newcombe, R. and Urry, S.A. (1983) *Construction Management in Practice*, Longman, Harlow.

Friedman, M. (1977) *Inflation and Unemployment*, p. 12, Institute of Economic Affairs, London.

Newcombe, R., Langford, D.A. and Fellows, R.F. (1990) *Construction Management 1: Organisation Systems*, Mitchell.

Part 2

Executing the Research

Chapter 3

Initial Research

The objectives of this chapter are to:

- introduce the **research process**;
- discuss the requirements for and methods of reviewing **theory and literature** critically and systematically;
- stress the importance of **assembling the theoretical framework**;
- emphasise the imperative of **proper referencing**.

The research process

In the 1980s, the Science and Engineering Research Council (SERC), the forerunner of the Engineering and Physical Sciences Research Council (EPSRC) in UK, held a Specially Promoted Programme (SPP) in Construction Management and issued the following diagram (Fig. 3.1) of their view of the research process relating to the SPP. Initial studies provide the foundation for all the research work that follows. Depending on the nature of the study, the initial work provides either the means for determining or confirming the aim, objectives and hypothesis or for confirming the topic for study. In either case, initial studies are essential to ensure that the research intended has not been carried out already and, more especially, they determine what has been researched and what issues are remaining or emerging for investigation. This helps to

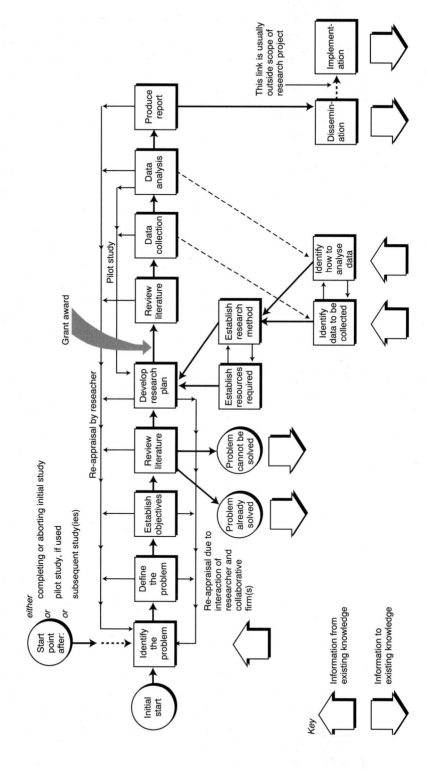

Fig. 3.1 Suggested construction management research process (source: SERC 1982).

avoid making the same mistakes that other researchers of the topic have made. Preliminary research involves searching sources of theory and previous studies to discover what the appropriate bases for the subsequent, detailed work are likely to be – often, alternatives will be found. It is at this stage that the design of the main research must be formulated or confirmed.

Research is a dynamic process. Therefore, it must be flexible – implying, although not requiring, that a contingency approach will be helpful. Early in the study, links between problems, which may be either topics or issues, theories, previous findings and methods will be postulated. The links should form a coherent chain, and so may need to be adapted as the work develops and findings emerge. The goal must be to maintain coherence and complementarity; only by such an approach will the results and conclusions be robust. The research path, outlined in Fig. 3.2, is embodied in the research design, data collection and data analyses, encompassing both the nature of the data and the methods used. Whatever method is adopted, it is essential that the research be conducted rigorously – that it is an objective and valid study.

Bechhofer (1974) considered the process of social research to be '...not a clear cut sequence of procedures following a neat pattern but a messy interaction between the conceptual and empirical world, deduction and induction occurring at the same time'. In examining the design of a research project, it is useful to consider the intended outputs from which the data collection requirements and the necessary analyses may be determined. Research may be regarded as an information system – the desired outputs are the starting point for determining what is necessary in the other main parts of the system, the inputs and conversion process, given the operating environment.

It is important to consider all the main processes of the information system – the desired outputs, the available data and information, the required conversions; the feedback mechanism makes 'checking' viable as well as allowing for development of the system; it should help to identify when and what environmental forces have an influence. In the dynamics of research, the process cycles through time, each new research project is able to build on those which have preceded it and *it is important* that they do so. Thus, it is essential that every researcher embarking on a project endeavours to discover

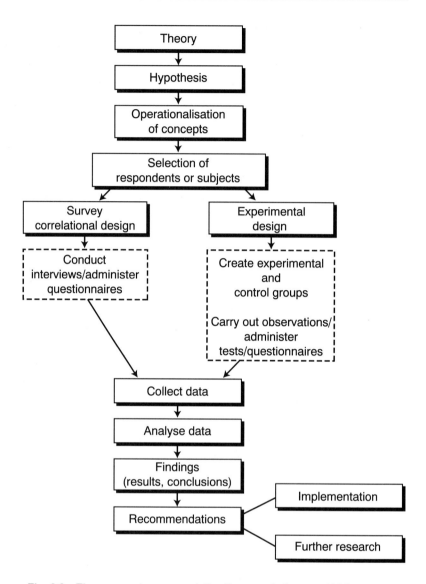

Fig. 3.2 The research process (after Bryman & Cramer 1994).

what relevant work has been executed, as well as what theory bases apply, otherwise the wheel may be invented repeatedly and, without a base of theory, there will be little understanding of what has been done and the foundation from which progress may be achieved. Hence, progress in the development of knowledge is likely to be constrained.

Considerations of theoretical bases and previous research will shed light on appropriate methodologies to aid replication of studies so as to approach the topic from alternative, but complementary, perspectives. Certain disciplines have traditions of employing particular methodologies and so, whilst a wealth of experience and expertise may have been accumulated, it may prove more difficult to establish the legitimacy of using a different methodology.

During the initial research phase, it will be useful to produce or, if produced already, to review the research model. Such a model will depict the main variables and the hypothesised relationships between them. Production of such a model begins at the conceptual level; that conceptual, or theoretical, model must be converted into an operational model – a model which can be used in practice to 'drive' the research and generate the variables which are to be observed and measured. In deriving the operational model from the conceptual one, inputs of previous research finding are employed – to determine what relationships (causalities) have been corroborated and which remain to be investigated.

Thus, the models identify what lies within the boundaries of the research project, known as the *endogenous variables*, and what lies outside the boundaries, called the *exogenous variables*. The 'permeability' of the boundaries determines the degree of influence exogenous variables may have on the system under investigation.

Data, from an information system perspective, may be regarded as 'raw facts and figures' – measurements which can be made and recorded. Information is facts and figures which are expressed in forms suitable to assist a decision maker; information directly supports decisions.

> *Example*
> A contractor's tender sum for a building project is an item of data; the presentation of that sum in the context of other tenders for the project with a discussion of the 'levels' of the prices bid is information; hence, usually, information is raw facts and figures which have been 'processed'.

Data stand alone whilst information incorporates data and places them in context(s) – data are objective whilst information often contains subjective elements. People's answers to questions in which they are required to express opinions is a form of subjective data. Provided the responses have been obtained properly (see Chapter 6), the responses do constitute data.

Theory and literature

An essential early stage of virtually all research is to search for and to examine potentially relevant theory and literature. Theory and literature are the results of previous research projects. Theory is the established principles and laws which have been found to hold, such as Einstein's theory of relativity; theories of the firm. Literature, in this context, concerns findings from research which have not attained the status of theory (principles and laws); often, it represents findings from research into particular applications of theory.

The items of theory and major references should be established in early discussion with the supervisor and others who are experts in the topic. Consultations to determine the usefulness of the proposal during its formulation will reveal appropriate theories and major research projects which have been carried out. These are good starting places, but it should be remembered that references are always historic although research journals, including leaflets published by research councils, private research organisations and professional institutions, often publicise current research projects. Employment advertisements for research posts often note the projects to be undertaken and so indicate where interest and expertise in those topics can be found.

Fortunately, despite work pressures, most researchers are keen to collaborate and help, but it is important that what is asked of such experts is well focused and shows reasonable knowledge of the subject and the desire to investigate a topic of some recognisable importance. A blanket request to 'tell me everything you know about... quickly and for free', will not be welcomed.

Research papers, which constitute the largest and the most important wealth of literature available, usually include a review of theory and literature which informed and underpinned the work reported, including the methods used. So, in proceeding to note data collected, the analyses executed, results obtained and conclusions drawn, research papers present distillations of previous work on the topic and advances made by that piece of research itself.

Assumptions and definitions

Reflect upon the topic. The draft proposal and model of variables almost certainly will contain terms to which particular meanings must be attributed.

> *Example*
> Often, industry contexts and meanings are *assumed* to apply and to be known and accepted widely – what is a 'frog'? In a bricklaying context it is a hollow (void) on one face of a brick; to the general public, it is a small, greenish amphibian which hops and croaks.

So, it is a good principle to identify assumptions and to define terms – management contracting and construction management are different in North America from the procurement routes given those names in UK. Literature is valuable in establishing the variety of terms and definitions, which are important points of debate, and hence one can select the definition most appropriate for use in the research project.

Before progressing, it is good practice to review the proposal with the supervisor of the research and with colleagues. The review should attempt to ensure that the assumptions and the important terms have been noted clearly and defined appropriately for the intended work. The test is whether the proposal can be understood unambiguously by any intelligent person; not just someone very familiar with the context and the particular topic of the proposed research.

Theory and literature review

The definition of the topic and terms must have been established during the production of the research proposal; the programme of work will show the time available, although it is usual for the review to be 'kept open' so that further literature can be incorporated if any work of significance is discovered during later stages of the project. However, care is required. Although keeping the review open in that way is useful to ensure it is comprehensive and to incorporate latest research findings, it may lead to the review never being finished – so it is important to establish a 'final deadline' to close entries to the review. Literature should not merely be found and reviewed, the body of relevant literature from previous research must be reviewed *critically*. The literature must be considered in the context of theory and other literature – the methodologies, data, analytic techniques, sampling etc. – so that objective evaluation takes place.

Neither in considering theory nor in the critical review of literature is it appropriate for the researcher to express personal opinion – let the theory and/or the literature 'do the work'. Alternative views and findings must be abstracted and ordered, to present thematic discussions such that a coherent debate is presented through synthesis and evaluation. Bodies of acceptance should be categorised as should arrays of issues – both of which inform subsequent portions of the research. Often, weighing of arguments in the context of theories and, if applicable, the theoretical stance adopted for the work is required. This provides and demonstrates appreciation and understanding of the state of knowledge of the topic and its context.

According to Haywood and Wragg (1982), the literature review must be critical and, therefore, demonstrate that '. . . the writer has studied existing work in the field with insight'. The insights should be derived from both the theoretical considerations and the completeness of the review of the literature. A mere listing of the articles which have been read with a summary of their main points is *not sufficient*; the critique – drawing out issues and arguments, setting alternative views against each other etc. – is the essential element of the *critical review*.

So, the review of theory and literature must provide the reader of the research report with a summary of the 'state of the art' – the extent of knowledge and the main issues regarding the topic which inform and provide rationale for the research which is being undertaken. Naturally, it is useful if the review not only informs the reader of the basics of the research, but makes the reader eager to read the subsequent sections of the report.

Conducting a search

A random search is unlikely to reveal much of significance for the topic – it is important that the search is structured. Often, it is best to begin with the theory on which the research is based – leading texts will be a good place to start as they provide statements and explanations of the theory and references to other work citations. Often, the references are useful in providing an introduction to the literature – seminal papers which, themselves, provide lists of further references.

As the scope of titles of publications can be enormous, even on a single, well-defined topic, it is helpful to have several approaches to discovering information, i.e. adopt a triangulated search. It is useful to list theories to be considered, leading authors and topic keywords. The more precise the search keys can be, the more both time and expense can be saved. Many 'on-line' searches are expensive and so, commonly, libraries restrict the number of words which can be entered for a search. It is a good idea to do some preliminary investigations to ensure that the best use of such facilities is being made.

Naturally, the lists of authors will grow and grow; that should not happen to the list of keywords – the list of keywords may change but it is a good idea to limit its size to a small number; many research journals (etc.) limit the number of keywords authors can use to a maximum of five or six, and a maximum of three is not unknown. So, *good topic definition is essential*.

Example

Consider writing a list of the libraries which it will be useful to consult or visit to obtain information for research.

Apart from containing the local university and college libraries, the list should contain other local libraries and specialist libraries – of professional institutions (CIOB, RICS, ICE etc.) and research organisations. Some large companies and consultants' practices have libraries of their own. The British Council local offices often have a limited library facility. More importantly, a link to the British Library (with its vast collection) can be available through the inter-library loan facility.

Obtaining access to libraries may present some difficulties, so it is best to enquire by telephone or letter before making a visit, especially if the library is some distance away. Most public, university and college libraries allow access for reading but may restrict borrowing. However, a charge, even for reading facilities, is made in some countries and it is useful to carry evidence of status as a research student of a university etc. to gain both access and assistance. Often, private libraries have more restricted access, such as to people with membership of the professional institution involved, although, by prior arrangement and evidence of the research status of the applicant, access can be obtained in most cases, it just may take a little time!

Usually, library staff are very helpful. However, an efficient search pattern is likely to involve consulting the local university subject librarian for initial advice and assistance and, once the primary information has been obtained and scrutinised, referring to specialist libraries for particular items and topics to complete the picture.

Increasingly, libraries employ electronic technology – computer data bases are replacing card indexes, which makes searches much quicker, easier and more comprehensive. It is helpful to consult abstracting services as a mechanism for preliminary selection of what papers to obtain – a title may not indicate the paper's contents accurately but an abstract, even if very brief, should provide a valuable synopsis of content and conclusions. Especially where particular researchers are 'key' to progressing in a topic (such as Milton Friedman and Monetarism), citation indexes are useful sources to trace developments in the topic via papers and researchers who have cited (made reference to) the 'key' individual. Clearly, the detective work in collecting literature involves a considerable amount of logical networking. *Do not lose sight of the boundaries of the topic.* Ensure what is collected is relevant to the aim, objectives and (if appropriate) hypothesis.

If not developed to some extent by this stage of the work, speed reading and writing are useful skills. Researchers need to reach a large quantity of material and abstract all potentially helpful items *with accurate references*. It may be helpful to provide the basic reference to the paper etc. at the top of a sheet and to note page numbers in the margin as brief quotations and summaries of contents are abstracted. As it may not be possible to borrow all the books or journals required, photocopying relevant sections may be possible – be careful in doing so that copyright is not infringed. Some sources may not be available immediately – books may be on loan, probably with a waiting list, copies of papers from other libraries etc. will take time to arrive. Whilst such time requirements may not be a major problem, they must be recognised and incorporated into the work programme. So, the earlier material is requested the better, but this possibility is dependent on how well the research topic has been defined.

When nearing completion of the collection of theory and literature or as the deadline for completion of that part of the research approaches, as indicated on the programme, the information should be categorised and ordered in a logical sequence to present the basis of the research in the report. Reference to the proposal which was produced will be helpful and, although alternative presentations are available, certain principles apply irrespective of which alternative is adopted.

As research builds on theory, it is helpful to provide a review of the theory before presenting the relevant literature of research findings. Theory and literature may be presented in a series of topic or sub-topic categories or the array of theory followed by the literature. To an extent, the issue is one of personal choice, but for larger projects involving a diversity of theory, it is useful to present theory and literature category by category with an overall summary review; this approach aids both flow of argument and understanding by a reader.

Example
Consider a research proposal on the topic of bidding for construction work (whether by contractors to win projects or by clients to let projects, including maintenance work etc.). Preliminary lists of aspects of theory and literature, including leading texts and research papers to be explored, might include:

Theory: Competition
Corporate objectives and strategies
Behaviour – individuals and organisations

Literature: Friedman (1956)
Gates (1967)
Carr (1977)
Hillebrandt & Cannon (1990)
Langford & Male (1991)

It is useful to keep the lists available for updating and extending as the initial research progresses so that, on completion, the lists of theory and literature in support of the research will be comprehensive.

Assembling the theoretical framework

Theory provides the framework for the research project rather like a structural steel or reinforced concrete frame is used in a building. It will also indicate the data which should be collected and further theory will denote appropriate methods and techniques of analysis.

It is essential that theories themselves be subject to rigour of analysis. Especially in social sciences, it is quite common to encounter theories which are in conflict. Principles and laws in these disciplines are derived from observations and analyses of human behaviour, so the complexity of integrating findings etc. leads to varying interpretations of behaviours and substantiations of those interpretations by further testing. The dynamism of societies complicates the research extensively – people's behaviours vary for many reasons; gravity and other physical and natural laws are reasonably invariate, whilst our understanding of such laws is developing still.

So, bodies of theory must be examined and evaluated to arrive at a theoretical basis or framework appropriate to the research proposed. It may not be possible to decide the logical body of theory to use from the description of theory provided, and it will not be possible to weigh alternative and possibly competing theories. In such a situation a choice must be made. The basis of such a choice may be personal preference caused by familiarity or expertise with an approach or set of ideas, a sympathy with the theoretical perspective (e.g. a Keynesian, a Monetarist or a Marxist view of inflation), or findings from leading research in the topic.

Of course, it is debatable whether competing theories can constitute basic 'principles and laws', or whether they are perspectives and beliefs which give rise to partly-supported hypotheses. Perhaps from a pragmatic perspective of a researcher using the theories and irrespective of what they are from a research philosopher's point of view, the 'true' nature of the theories is relatively incidental *provided* their natures are recognised and taken into account in the execution of the research project. The question is one of academic contention – especially in terms of the philosophy of research and the validity of findings; what are realities, what are truths and how do they differ? At this stage, it is important and sufficient to be aware of issues and to recognise and use theories for what they are – limitations and all.

So, the theory adopted provides the basic structural framework to identify and explain facts and the relationships between them. Consider a diagram of the variables in a research proposal: determination of the variables, identities of the variables, and the relationships between them should be determined from theory. Hypotheses are employed to fill in gaps – to suggest relationships which may exist if theory is 'extended', or to relate two aspects of theories. Where theories are really partly-supported hypotheses, those hypotheses are appropriate for further testing, so for example, it is likely to be appropriate to investigate which theory of inflation is more appropriate in a particular situation.

It is a good idea, whenever possible, to use theory to build a model of the proposed research – the variables and relationships, the points of issue and those of substantiation. In cases where assumptions are involved, maybe as a requirement or pre-condition for the theory or law to hold, such as Boyle's Law, it is essential to note the assumptions *explicitly* as they impose limitations on the research and its findings as well as prompting questioning and, perhaps, investigation of what occurs when the assumptions are relaxed. The main things to be done and their sequence, to assemble the review of theory and literature into a theoretical framework, include:

- Defining the topic and terms; time and cost limitations.
- Noting items of theory.
- Noting major references known.
- Listing keywords for theory and the topic.
- Locating libraries.
- Obtaining access to libraries.
- Investigating library resources – abstract and citation indexes, computer search facilities etc.
- Executing searches.
- Obtaining and reviewing sources with concise recording of information.
- Assembling the review.

Proper referencing

If sections of the work are not referenced, they are assumed to be original work of the researcher. If that is not the case and the work, ideas, etc. have been obtained from someone else, presenting that work as the researcher's own is *plagiarism* – intellectual theft – and it is treated very seriously indeed. However, virtually everyone makes mistakes and occasionally omits references accidentally. Plagiarism is omitting references deliberately; it has been found to the extent of copying and submitting someone else's entire dissertation. In such instances condemnation will be swift and complete.

There are several standard methods for referencing. The Harvard system is used very widely and is gaining in popularity. Most publications and institutions prescribe the referencing system to be used.

Example
The Harvard system of referencing is:

Textbook:

In the text of a report:

Quotation: 'What was the relationship between theory, data collection and analysis?' (Burgess, 1984)

Paraphrase: Burgess (1984) posed the question of the relationship between theory, data collection and analysis of those data.

In the references section of the report:

Burgess, R.G. (ed.) (1984) *The Research Process in Educational Settings: Ten Case Studies*, Lewes, Falmer Press.

Journal:

The entries in the text of the report are as shown for textbook.

In the references section of the report:

Roth, J. (1974) 'Turning adversity into account', *Urban Life and Culture*, **3**(3), 347–359.

Summary

In this chapter, we have considered the requirements during the early stages of carrying out the research work – the collection and reviewing of theory and literature. Mechanisms to assist collection of such information have been discussed; in particular, the necessity is to be systematic and rigorous. It is essential that the review produced in the research report is a critical review and is referenced correctly.

References

Bechhofer, F. (1974) Current approaches to empirical research: some central ideas. In: *Approaches to Sociology: An Introduction to Major Trends in British Sociology* (ed. J. Rex), Routledge & Kegan Paul, London.

Bryman, A. & Cramer, D. (1994) *Quantitative Data Analysis for Social Scientists* (revised edn.), Routledge, London.

Carr, R.I. (1977) Paying the Price for Construction Risk. *Journal of the Construction Division, Proceedings of the American Society of Civil Engineers,* CO1, 103, March, pp. 153–161.

Friedman, L. (1956) A Competitive Bidding Strategy. *Operations Research,* 4, pp. 104–112.

Gates, M. (1967) Bidding Strategies and Probabilities. *Journal of the Construction Division, Proceedings of the American Society of Civil Engineers,* CO1, 93, March, pp. 75–107.

Haywood, P. & Wragg, E.C. (1982) *Evaluating the Literature,* Rediguide 2, School of Education, University of Nottingham, Nottingham.

Hillebrandt, P.M. & Cannon, J. (1990) *The Modern Construction Firm,* Macmillan, Basingstoke.

Langford, D.A. & Male, S.P. (1991) *Strategic Management in Construction,* Gower, Aldershot.

SERC (1982) Specially promoted programme in construction management – *Science and Engineering Research Council Newsletter,* Summer, SERC.

Chapter 4

Approaches to Empirical Work

> The objectives of this chapter are to:
>
> - consider the **role of 'experience'**;
> - examine the requirements of **modelling** and **simulation**;
> - examine the requirements of **experimental design**;
> - discuss **qualitative** and **quantitative approaches**;
> - discuss issues of **collecting data from respondents**.

Role of experience

When does research begin?

It is all too common for people to believe that research has not really started until the data collection has begun. That is wrong. A problem which occurs very frequently is that the data collection is begun prematurely – before the theory and literature has been reviewed and, in extreme cases, before the proposal has been finalised. It is hardly surprising that, in such cases, two significant problems arise concerning the data. The problems are collecting data which are not relevant to the research, and failing to collect data which are necessary. There may be further difficulties over the size and nature of the samples. It is important to remember that it is difficult enough to collect data once; having to collect a second, supplementary set is compounding the difficulty. A researcher and, by implication, other researchers, will lose credibility by returning to respondents in order to remedy gaps in data collected. The target of obtaining high quality data is applicable – get it right first time, you are unlikely to get a second chance.

What is experience?

Strictly, empirical work is concerned with experience gained from experimentation. This leads to the question of 'what is experience'? If experience is a form of human learning, by definition, it involves observation, evaluation, memory and recall. All four activities include problems of selection and accuracy, so experience is unlikely to be totally reliable – some observations are missed, others are interpreted and understood incorrectly, memory can distort and recall may lead to omissions, that is, memory is selective and deficient.

Usually, people are blamed more frequently than they are praised and remember blamings much more vividly. Not only does that demonstrate selective recall etc. but also has implications for behaviour both of the individual and of other people. 'Bosses' tend to blame more readily and more frequently than they praise. The consequence of such behaviour, the more vivid recall of blamings and the selectivity of memory, means that experience induces people to focus on avoidance of repeating mistakes. This may invoke conservative behaviour – perpetuation of the *status quo* in terms of performance etc.

What is important for research is that there are problems of comprehensiveness and accuracy in relying on experiences – hence, the necessity to record all data accurately and speedily, irrespective of the research methods adopted; this necessity is well known and practised rigorously by those concerned with laboratory-based, experimental research.

Modelling

What is modelling?

Modelling is the process of constructing a model, a representation of a designed or actual object, process or system, a representation of a

reality. A model must capture and represent the reality being modelled as closely as is practical, it must include the essential features of
the reality whilst being reasonably cheap to construct and operate
and easy to use.

> *Example*
> Models occur in a variety of forms and serve many purposes. A toy
> car, which must resemble a tiny version of the real car which it
> represents, enables a child to learn through play – often play which
> simulates the actions of a real car in situations both experienced and
> imagined by the child. Toy building bricks and architectural models
> offer different levels of sophistication (detail, complexity, accuracy) in
> representing buildings, whilst mathematical models are employed by
> engineers in the design of structural components and building
> services systems. Economic and econometric models are used
> extensively – project cash flow models, models of resource inputs
> required for different types of project, the Treasury's models of the
> UK economy. Population models are used to forecast demand in
> different sectors of the economy. Another main function of models is
> to facilitate reasonably accurate prediction.

Classification of models

Mihram (1972) discusses some alternative classification of models –
those of Rosenblueth and Weiner (1945), Churchman *et al.* (1957)
and Sayre and Crosson (1963).

Rosenblueth and Weiner (1945) categorise models in science as:

- material models: transformations of original physical objects,
- formal models: logical, symbolic assertions of situations,
                     the assertions' representing the structural
                     properties of the original, factual system.

Subcategories, which alternatively may be regarded as alternative
categories, are:

- open-box models: predictive models for which, given all
                     inputs, the outputs may be determined,

- closed-box models: investigative models, designed to develop understanding of the actual system's output under different input conditions.

Churchman *et al.* (1957) suggested that models fit into the following categories:

- iconic: visual or pictorial representation of certain aspects of a real system, such as computer screen icons to denote programmes; detail drawings of parts of a building,

- analogue: employs one set of properties to represent some other set of properties which the system possesses, (e.g. electrical circuit to mimic heat flow through a cavity wall)

- symbolic : requires logical or mathematical operations (e.g. equation of an 'S curve' of project cash flow).

Sayre and Crosson (1963) suggest the categories of:

- replications: display significant physical similarity to the reality, such as a doll,

- formalisations: symbolic models in which more of the physical characteristics of the reality are reproduced in the model; symbols are manipulated by techniques of a well founded discipline such as mathematics (e.g. $y = a + bx$ is the equation of a straight line),

- simulations: a formalisation model but without entire manipulation of the model by the discipline's techniques in order to yield an analytic solution or a numerical value (e.g. construction project bidding models).

Distillation of the various classifications of models suggests that the common forms of models are:

- iconic,
- replications,
- analogues,
- symbolic.

For research purposes, the more common forms of model are analogue and symbolic, whilst in the construction industry iconic models and replications are usual.

Deterministic and stochastic models

All models contain parameters (variables) which must be identified and quantified for use in the model, together with their inter-relationships. The resultant models are either deterministic – what happened in the past will be replicated in the future – or stochastic (probabilistic) – the laws of probability which governed past realisations will continue to apply in the future. Deterministic models tend to be simpler in form and in the manipulations required than their stochastic counterparts. Whilst, by definition, stochastic models cannot take account of 'shocks' which may occur in reality to the system under study, they are likely to be more realistic but more complex representations. Shocks produce discontinuities in the operation of models – such as changing the 'levels' of inflation forecast by economic models when the oil crises of 1973/4 and 1979 occurred.

Deterministic models assume away random effects, whilst stochastic models seek to incorporate them. A stochastic model will either 'simulate random variables for a whole range of statistical distributions' (Morgan 1984, p. 5) or simulate the particular distribution, if known. For a discussion on the assumptions about distributions and their use in construction project bidding models and risk management (see Chau 1995, 1997 and Fellows 1996). Whether deterministic or stochastic, a model should mimic the effects present in the reality by inclusion in the constituents of the model. For deterministic models, this may be an 'express' residual element which is not included in any other component. For stochastic models it may mimic reality by incorporation of probabilities, whether or not a distribution type is assumed or determined.

For realities in which changes occur only slowly or consistently, deterministic models can be appropriate – the pattern of any consistent change can be determined and incorporated to yield dynamism to an otherwise static model, such as in the deterministic analysis of time series discussed in Chapter 7. Similarly, dynamism can be incorporated in stochastic models. Dynamic components of models may be continuous, as in a growth trend, or discrete, as in seasonal elements of construction workload. 'Most systems... develop their characteristic behaviour over time, so that any... models of this behaviour needs to be... dynamic...' (Mihram 1972).

The modelling process

Models may be used to investigate and/or to predict; for managers, predictive models are more valuable, whilst auditing requires investigative modelling. PERT is a stochastic predictive model; an investigative model could comprise a set of equations in several unknowns, sufficient that provided a certain number of values of some of the variables are known, the equations can be used to determine the remainder, as in linear programming. The modelling process is depicted in Fig. 4.1.

The objective of the model should be to reflect the purpose of the model, such as the questions to which answers are to be sought from the model. One should know for whom the model is to be constructed, in order to lend perspective to the modelling and to suggest sources of data, forms of outputs etc. The analysis stage comprises organised, analytic procedures to determine the operation of the reality, noting the location and permeability of the boundary of the system to be modelled. Often, a diagram of the reality will be of benefit in identifying variables and their relationships prior to the quantification of both. This is a major element of the synthesis stage, which yields one model or an array of alternative models. It is likely that the resulting models will reflect both the education and training of the analysts and their experiences of modelling other, especially similar, realities.

Verification of a model involves determining whether the structure of the model is correct; this is achieved by testing the model, by examining the outputs resulting from the model under a given set

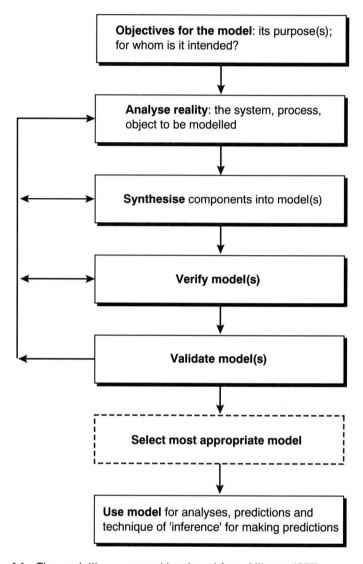

Fig. 4.1 The modelling process (developed from Mihram 1972).

of inputs. The model is verified if the outputs are appropriate, i.e. they approximate to 'expectations' of what a good model would yield. Models which are verified may pass to the next stage, *validation*, whilst those which are not verified may be discarded or, especially if only one model is being examined, be returned to the analysis or synthesis stages for further scrutiny and amendment.

In validating a model, the model's output resulting from known inputs is compared to realisations of the reality, such as an ex-post forecast (see Fig. 5.1). If possible, it is helpful to carry out such validatory testing for several sets of inputs and known outputs of the reality to examine consistency of the model over a range of conditions, preferably including 'extremes'. If a number of models have been suitably verified, it is usual for validation to be used to select the most appropriate models. Verification may suggest a model which is 'best' on the basis of theoretical 'fitting' – from criteria based on analysis of the model, whilst validation may yield a different 'best' model on the basis of closest fit of output to test realisations. The choice of model will depend on the objective of the modelling exercise, its use and by whom it will be used etc., and the differences in the two forms of performance between the models being tested. In the final stage, the verified and validated model may be used. However, the model also may be iterated so that inferences can be considered about the possibility of extending the operation of the model to other conditions, by relaxing some of the restrictions or assumptions for the models' appropriate operation and use.

The formulation and construction of a model requires a variety of inputs, as does any research activity. Following the decision of the objectives and limitations of the research, the initial stage is to investigate existing theory and principles. Once appropriate theories and principles have been distilled from the existing body of knowledge, literature can be searched to determine the applications of these theories and principles and findings thereof, in research projects and in practice. Such investigation will indicate appropriate variables to define, isolate and measure (usually via experimentation, whether in a laboratory environment or in-use in a 'reality'), so that performances of the individual variables and their relationships can be evaluated.

Once the structure of the model has been established and its performance scrutinised and determined to be suitable for the objectives, appropriate values can be input for the necessary variables and the resultant outputs calculated. Clearly, the direction of using a model may not be the same as was employed to construct, verify and validate the model.

> *Example*
> In structural engineering, the performance of a member may be modelled by constructing various test members (for experimentation) of different sizes and combinations of component of known properties (tensile strength of steel, compressive strength of concrete etc.) to establish its load-bearing capabilities, performance characteristics and failure modes. The resultant model then may be used (with appropriate 'factors of safety') to design the components required for the member to achieve the necessary performance characteristics.
>
> The example of Time Series Analysis and Forecasting, in Chapter 7, may be regarded as an instance of elementary modelling.

Simulation

Dynamism

Simulation involves the use of a model to represent the essential characteristics of a reality, either a system or a process. So, whilst a model may be a static representation, such as an architectural model, a simulation involves some element of dynamism, if only because it models a process rather than an object. Thus, flight simulators mimic the behaviours of aircraft under specified conditions; models in wind tunnels are used to investigate the flow of air (simulations with scaling) about the model. Hence, often, simulation is used to examine how the behaviour of the reality is likely to change consequent upon a change in the values of input variables etc. in the representative model.

Simulation is used to assist prediction of the behaviour of a reality or/and to revise a model to enhance its predictive accuracy or predictive capability. Morgan (1984) suggests a variety of purposes for simulation:

- explicitly mimic the behaviour of a model
- examine the performance of alternative techniques
- check complex mathematical/analytic models
- evaluate the behaviour of complex random variables, the precise distribution(s) of which is (are) unknown.

Heuristics

Clearly, as modelling lies at the base of simulation, formally in many instances but informally in others, the procedures employed to obtain an appropriate model of a reality apply to simulations. Increasingly, computers are used in simulations; often, such simulations use *heuristics*, or 'rules of thumb' in their replicating behaviour of a reality as many realities comprise a complex system of independent but interactive components which act together. Neither the individual components and their behaviours, nor the interactions of those components may be known, understood and modelled in detail or exactly, so, if only due to practical constraints such as time, cost etc., a sufficiently accurate simulator may be produced through observation and measurement of the reality. Through interpretation and interpolation, deduction and induction and given knowledge of appropriate theory and principles, heuristics for the simulator may be produced.

Mitchell (1969) notes that, 'in a practical context . . . [simulation can] . . . be an extremely powerful way of understanding and modelling a system. More importantly, from the practical point of view, it can be the only way of deriving an adequate model of a system'. Meier *et al.* (1969) extend the assertion, stating that 'experimentation', by means of computer simulation, can overcome some of the restrictions that exist when other forms of analysis are used. By using representative heuristics, simulation, 'opens up the possibility of dealing with the dynamics of the process, too complex to be represented by more rigid mathematical models . . . . Simulation may make possible experiments to validate theoretical predictions of behaviour in cases where experimentation on the system under study would be impossible, prohibitively expensive or complicated by the effect of interaction of the observer with the system . . . '.

Approaches

Whilst simulation may be used to represent the behaviour of a precise model in a realistic way, because it is a model in which the natures or distributions of the dynamic elements are known, as for parametric statistics, simulation also may be used to evaluate the behaviour of a system of random variables of which the distributions are unknown as for non-parametric statistics. In the latter case, theoretical, experimental and experiential evidence may be used to suggest appropriate distributions to be used in simulations, although considerable debate over the appropriate distributions may ensue (see Chau 1995, Fellows 1996, Chau 1997).

Example

Frequently, the Monte Carlo technique of random number selection is used in simulations. About 1000 iterations is normal in computer simulations employing Monte Carlo for predicting durations, costs and prices of construction activities.

Whilst the form of the assumed distribution of the random variable may influence the result of the simulation, it is acknowledged that the limiting values used for the range of possible values which the variable may take is important. Such a view is significant in PERT's use of β-distribution and the determination of each activity's limiting values, optimistic and pessimistic, which may be determined by many techniques – Delphi, analyses of past performance, expert estimation etc.

Game approaches are popular as simulations for people making decisions. Business games are used to enable players to see the consequences of decisions made in a simulation of an organisation operating in an environment, using heuristics as its operating rules – for both environmental and organisational changes and outcomes. Clearly, learning by the players is a major objective, but games may be used in similar ways to replicate organisations and to make predictions – such as for alternative 'futures'.

Example

A major gaming simulator for operating a construction company is AROUSAL (developed by Peter Lansley, University of Reading, UK) which provides iterations of bidding, resourcing projects and the company over a pre-determined period of the company's operation; projects available to bid are offered, results provided, accounts for the projects and company produced, personnel profiles for employment and of performances are available periodically etc. Such gaming allows managers to practise and develop skills in a realistic but 'safe' context.

Generally, games have expressed or assumed objectives, and sometimes both, and they have rules, but the outcomes are determined by inputs and by circumstances, which are random variables. Hence, the need to appreciate risks, at least as outcome variabilities are demonstrated, plus how and to what extent they may be influenced. *Partnering* is cited as a win–win game because, if the participants, partners or players adhere to the rules of behaviour, all will benefit. Negotiating a final account under 'traditional' project arrangements is a zero-sum game, since what one party gains, the other must lose and because the aggregate is a constant, it is the pattern of sharing that aggregate which is at stake.

Simulation offers a unique opportunity to observe the dynamic behaviour of complex, interactive systems. A carefully constructed, realistic simulation model provides a laboratory environment in which to make observations under controlled conditions for testing hypotheses, decision rules and alternate systems of operation under a variety of assumed circumstances.

Experimental design

Defining an experiment

An experiment is an activity or *process*, a combination of activities, which produces *events*, possible outcomes. Usually, in scientific contexts, experiments are devised and conducted as tests to investigate any relationship(s) between the activities carried out and the resultant outcomes. Tossing a coin a number of times could be used as an experiment to test for bias in the coin; likewise, the throwing of a die. Hicks (1982) defines an experiment as a 'study in which certain independent variables are manipulated, their effect on one or more dependent variables is determined and the levels of these independent variables are assigned at random to the experimental units in the study'.

Ideally, variables should be isolated through the design of an experiment such that only one of the, possibly, very many independent variables' values is changed and the consequences on the isolated single dependent variable is monitored and measured accurately.

Hick's definition raises the issue of the way in which the independent variable is 'manipulated'; although random variation is one approach, commonly, particular values within a 'range of interest' are assigned to the independent variable. This method provides practicality, but also some restriction on the inferences which can be drawn from the results. In social investigations, including construction management and construction project-based experiments, it is neither practical nor possible to allow only one independent variable to alter in value, nor is it possible to isolate individual dependent variables on most occasions, hence the usual approach to experimental design is to devise a study in which the main independent variables, except the one of interest, are held *approximately* constant and the consequences for the major dependent variable are measured. A common approach is to undertake comparative studies on similar projects executed at about the same time by similar firms employing similar organisational arrangements. Such a study could investigate the impact of different management styles of project managers on project performance, measured values of time, cost, quality, etc. Variables of location, weather etc. as well as (preferably, minor) differences between the 'common independent variables' should be acknowledged in the evaluation of results.

An experiment is designed and it occurs in the future. However, there are many instances where analysis is required of data which have been collected in the past; such an approach cannot be an experiment but is known as *ex-post-facto* research. Ex-post-facto research is very useful in many contexts, such as building economic models from which forecasts can be made. In testing such models, it is essential to ensure that the data used for the test have not been used in the model-building (see sections on modelling and simulation).

Commonly, it is believed that experiments and their designs must begin with a statement of the problem or issue to be investigated. However, that requirement does not apply to experiments and, further, is not the real initiation of research; that initiation is recognition that a problem/issue may exist, and which gives rise to the question, 'what is the research intended to find out?'. Given that the question is asked, it must be answered as precisely as possible in order that a statement of the intended investigation can be made, noting requirements, parameters and limitations. In this way, the most appropriate methods by which to carry out the research can be determined. Thus, given resolution of the preliminary issues, the devising and design of an experiment begins with, and is driven by, the statement of the problem.

Variables

In common with any information production, the objective is to support and facilitate decision making; in the case of experimental research, the decision making concerns inference about the relationships investigated. Hence, the variables must be identified, together with appropriate definitions and measures; assumptions should be explicit so that appropriate hypotheses for testing by the experimentation can be formulated expressly. Commonly, such hypotheses concern relationships between independent and dependent variables, to assist examinations of strength and direction of the relationships and, in view of theory and literature, causality.

Clearly, measurement of the variables is crucial. Experimental design considers the degree of accuracy which can be achieved and the method of achievement. Given that no measurement will be 100% accurate, the criterion is to obtain sufficient accuracy for the purpose of the experiment; if alternative experimental or measurement techniques are available, the most accurate means, subject to the pertinent constraints, should be selected. For example, in forecasting accuracy of an experiment, the probability of errors etc. should be considered.

A major consideration in designing an experiment is the method used to change the independent variable in order that any consequential changes in the dependent variable can be measured. Three main approaches to effecting the changes in the independent variable are employed:

- **Randomised change** – of the independent variables and/or their values – perhaps by use of some random number generator to determine the values to be employed, within or without limits to the 'range of interest'. Randomisation allows the experiment to be conducted, results produced and conclusions drawn, using the common assumption of independence of errors in measurement; this is usual in much statistical analysis – randomisation validates the assumption. Further, randomisation supports the assumption of 'averaging out' the effects of uncontrolled independent or intervening variables. Such averaging removes much of the effects of the uncontrolled variables but does *not* do so totally – those variables increase the variance of the measured values of the dependent variable. Randomisation is helpful in eliminating bias; it ensures that no variables, or their possible values, are favoured or disregarded as there is no 'systematic' selection of either variables or their values. Further, it ensures independence between observations and measurements, a requirement for validity of significance tests and interval estimates.

- **Selected ranges of variables** – both in terms of the identities of independent variables and the ranges of the values which they may assume. For such experiments, the main independent variables must be identified by scrutiny of theory and literature etc. The variables can be quantitative and/or qualitative and their values selected or random. Use of extreme values of the variables should result in maximum effects on the dependent variables and, hence, both the identification of the range of consequences which might result, and a chance to focus on the most likely outcomes.

- The most restrictive but, often, the most convenient/appropriate approach, is to **control the independent variables rigidly** – the variables are determined and values assigned over the duration of the experiment (as in Boyle's Law experiments etc.). Strictly, the inferences which may be drawn from such experiments are valid for the fixed experimental conditions only.

Given more than one independent variable of interest, a factorial experimental design results. If there are two independent variables, α, β, where α may assume (for the purposes of the experiment) 5 values and β may assume 3 values, the result is a 5×3 matrix such that 15 combinations of the values of the independent variables must be investigated.

Replication

A universal, desirable feature of research, notably experiments, is replication – hence, it is essential to make meticulous notes of record of *all* events in detailing the conduct of an experiment. Good laboratory practice involves detailed and precise recording of all occurrences during the conduct of an experiment for subsequent scrutiny. Without complete detail of experiments, replication is not possible. As replication facilitates increased numbers of observations and measurements of the variables under *identical treatments*, it assists provision of an estimate of experimental error, and identification and quantification of sources of error. Further, it should lead to reduction in standard errors, thereby increasing precision:

$$S_{\bar{Y}} = \sqrt{\frac{s^2}{n}}$$

where

$S_{\bar{Y}}$ = standard error of the mean (of the dependent variable, Y)
s^2 = sample variance
n = number of observations.

The result is that replication assists inference; as replication increases, so a wider variety of situations and variables etc. can be subject to the experiment, thereby yielding a greater range of conditions to which the results apply, and so the inference base is broadened.

Petersen (1985) suggests that, due to the nature of standard errors, the accuracy of experimental results can be improved by:

- Increasing the size of the experiment by
  - replication,
  - incorporating more 'treatments' (values of independent variables).
- Refining the experimental method or technique to achieve reduction of experimental error through reducing the sample variance (s^2).
- Measuring a *concomitant variable*, another, associated independent variable, to facilitate *covariance analysis*, the variance of the combination of the variables, this may yield reduced experimental error.

In considering factorial experiments, described below, Petersen notes the following, additional considerations:

- Increasing the size of the experiment (as above) can be self-defeating if it requires the incorporation of more heterogeneous experimental units.
- Accuracy can be increased through selection of treatments such that the factorial combinations include hidden replication for some comparisons.
- Arranging experimental units into homogeneous groups can remove differences between groups. This reduces sample variance (s^2) and hence, experimental error.

In examining the results of any experiment, it is likely to be helpful to carry out an analysis of variance (often identified as ANOVA or, for multivariate analyses, MANOVA, in computer statistics programs). Analysis of variance is a systematic approach which identifies the constituents of the total variation, and thereby apportions the total variation to the contributing sources. Such analysis can be very helpful in refining experimental designs.

Randomised experiments

This is the basic design in which values of variables are allocated at random. Much of statistics employs the assumption of such randomness and the statistical analysis tends to be quite straight-forward. Randomised design offers greatest flexibility; however, its precision can be low, especially if experimental units are not uniform (Petersen 1985).

Randomised groups

The experimental groups or blocks are composed of units which are as near as possible homogeneous. This may be achieved by random allocation of units to groups or by precise design allocation of the units to groups. The aim is to avoid differences between the groups, thereby increasing precision by eliminating inter-group (between group) differences from experimental error. Thus, to be effective, the intra-group (within group) variance must be much smaller than the variance over the entire set of units. Hence, Petersen notes that:

- the size of each group should be as small as possible because, usually, precision decreases as the size of the group increases.
- If there are no obvious criteria for designing the group, 'square' group designs are most appropriate.
- If 'gradients' apply to groups (e.g. slope, age, strength), groups should be designed to be narrow and long (rectangular and perpendicular to the gradient).

Given Z treatments, each replicated n times, results in the need for nZ experimental units. The units should be placed into n groups, each of Z units, in a way so that the groups are as similar as possible. Then, the treatments are assigned to the units randomly such that each treatment occurs only once within each group. A notable problem in group designs is that missing data can cause problems for analysis.

The use of a randomised group design can increase the information obtained from an experiment, as the groups can be at different locations and the individual elements of the experiment can be carried out at different times. Thus, sampling can occur over a wider variety of circumstances. The aim is to separate the effects of the treatments from uncontrolled variations amongst the experimental units or groups. The treatments, units and their grouping and observations and measurements should be designed such that:

- Experimental units which are subject to different treatments should differ in no systematic way from each other (i.e. they should be unbiased; differences should be random, if any, and small).
- Experimental error should be minimised and achieved by use of as few experimental units in each group as possible.
- Conclusions should have maximum validity (concerning breadth and depth).
- Experimental technique should be as simple as possible, commensurate with the objectives (i.e. *parsimonious*).

- Statistical analyses of the results should not require the use of assumptions which are too restrictive or are inappropriate in the context of the objectives and the circumstances of the experiment.

Factorial experiments

Factorial experiments can be considered as constrained instances of randomised groups. For convenience, the main independent variables are identified and assigned values to be considered towards the extremes of their practical or usual range. Such an approach limits the number of combinations of continuous variables to be analysed. This is due to 'central limit' type of effects and the opportunity to interpolate between the results of the experiments so undertaken.

Example
To consider ready-mixed concrete from two suppliers, variables such as water–cement ratios, maximum aggregate size and amount of plasticiser could be assigned high and low values to yield a $2 \times 2 \times 2 \times 2 = 16$ 'cell' experiment, i.e. 16 possible combinations of factors affecting the strength achieved.

Note: Increasing the number of independent variables to be examined would extend the number of experiments considerably. However, the technique of 'Latin Squares' can be employed to restrict the number required and to maintain significance of and confidence in the results (see Levin and Rubin (1991) pp. 286–287).

Qualitative approaches

When are qualitative approaches employed?

For a number of years, the scientific method, with its emphasis on quantitative studies, has been in the ascendant, with a result that research in disciplines which lie between the natural sciences and social sciences, notably management of technology and engineering, has been drawn or pushed towards adoption of quantitative scientific method. However, quite recently, increasing recognition of the value and appropriateness of qualitative studies has emerged. This may perhaps be in acknowledgement of the potential for such methodologies to get beneath the manifestations of problems and issues which are the subject of quantitative studies, and thereby, to facilitate appreciation and understanding of basic causes and principles, notably, behaviours.

Tesch (1991) identified three categories of approach to the analysis of qualitative data:

- Language based – focuses on how language is used and what it means – discourse analysis, ethnomethodology and symbolic interactionism.
- Descriptive or Interpretive – attempts to develop a coherent and comprehensive view of the subject material from the perspective of those who are being researched; the participators, respondents or subjects.
- Theory-building – seeks to develop theory out of the data collected during the study; *grounded-theory* is the best known example of this approach.

Oakley (1994) suggests that the word 'qualitative' is used to describe research which emerges from observation of participants. She asserts that such research has two sources:

- Social anthropology and
- Sociology.

Sociological studies often were conducted on Westerners by Westerners which enabled the researcher, as a member of the population under study, to use knowledge of the society to isolate themes and to prepare frameworks within which study of a particular aspect could proceed. As anthropological studies were carried out by Westerners on non-Western societies, the work had to be more 'open'. The researcher could not have valid preconceptions of the society and so, not being part of it and having no initial understanding or knowledge of it, could not isolate themes or provide a framework for restricting the scope of the study. Hence, all data had to be captured and examined to enable hypotheses and theories to emerge.

Whilst sociologists were able to employ questionnaires and interviews, because they are influenced by the cultures in which they are devised and conducted, the more basic, anthropological studies could not employ such techniques. All studies involving people are influenced by cultures; to a degree, that must include all studies.

Example
The topic of *indexicality* notes that a person's understanding of a term etc. is dependent upon cultural and contextual factors. Such a consideration is important when examining conditions of contract, perhaps for the purpose of researching disputes and claims in the construction industry (see, for example, Clegg (1992)). Clearly, it can be a very important and problematic consideration when drafting a contract for use internationally.

A consequence of considering the nature of the subject to be researched is that it may not be possible to isolate a particular, defined topic to study – what can be studied emerges from the research through what is observed. In such cases, it is not possible to develop an hypothesis to test. Furthermore, the aims and objectives are likely to be framed loosely and be quite 'open-ended'. Commonly, the subject of culture arises in debate over how members of an industry behave in various circumstances – whether concerning observations of behaviour or predictions. For example, it is said that the construction industry has 'a macho culture' and that it has 'a culture of conflict'. Usually, such expressions are the result of casual observations and are influenced by the values and experiences of the observer. In fact, very little research has been done to determine the cultures of the industry – the values and beliefs which govern people's behaviours.

Example
Consider researching the culture of engineers. As a non-engineer, and given appropriate definitions of terms, the research should first investigate codes of conduct etc. and formalisations of expected behaviours of engineers and then proceed to observe how they operate in order to devise hypotheses over cultural factors. Those hypotheses may be tested subsequently through case studies, interviews, questionnaires etc.

Of course, other different approaches could be employed. The findings may be common, different or a mix but, irrespective of that, their validities and applicabilities will vary, dependent upon how the studies have been carried out. Execution of such open-ended studies requires not only meticulous recording of all data, but constant scrutiny of the data to aid the recognition of themes – variables and patterns of relationships between them.

Development of theory from data

The strategies of the open-ended research approach are encapsulated in *grounded theory* – see Glaser and Strauss (1967) – which involves the discovery of theory from data. The technique involves the gathering of data from observation of the sample. Next, the researcher examines the data from the perspective of the issues to be investigated through the research, and identifies categories of the data. Further collection of data follows until, with continual examination of the data and the categories, the researcher is satisfied that the categories are suitable, that they are meaningful and important – that they are *saturated*.

Further work can be undertaken to investigate generalisation of the categories and links between them. Such work will employ hypotheses to be tested by additional field work with a new or extended sample and may employ the technique of *analytic induction*. Analytic induction is a step-by-step process of iteration and evaluation. Initially, the issue is defined (perhaps only approximately), instances of the issue are examined and potential explanations and relationships are developed. Further instances or samples are investigated to analyse how well the hypothesised explanations and relationships apply. Such iterations continue until the hypotheses suit what is found in the data at a suitable level of statistical significance in appropriate cases.

In using grounded theory and analytic induction, Strauss (1987) emphasises that researchers must be, '...fully aware of themselves as instruments for developing that grounded theory.' The statement means that the researcher must be rigorous and highly objective in analysing the data to yield the categories and, thence, theory. In instances where hypotheses have been developed for testing, any desires to support or refute the hypothesis by the researcher must be ignored. In the research process, to ensure accuracy and validity, the research must *avoid bias*.

In carrying out studies which require the development of theory from data and the subsequent testing of the theory, where data collection, analysis and development of theory proceed together iteratively, Schatzman and Strauss (1973) advocate segregation of the researcher's field notes into:

- Observational Notes (ON),
- Theoretical Notes (TN) and
- Methodological Notes (MN).

Observational notes (ON) concern the recording of '...events experienced principally through watching and listening. They contain as little interpretation as possible and are as reliable as the observer can construct them.' Theoretical notes (TN) are 'self-conscious, controlled attempts to derive meaning from any one of several observation notes'. Methodological notes (MN) concern how the field work is carried out, and record any necessary changes, the reasons for such changes and when the changes occurred. Irrespective of the research methodology adopted for any project, taking detailed field (laboratory) notes is vital. The categorisation advocated by Schatzman and Strauss is appropriate for any research project, whether using qualitative or quantitative methods.

Much qualitative research concerns the generation of concepts through the researcher getting immersed in the data collected in order to discover any patterns. In so doing, it is essential to be sensitive in order to detect inconsistencies and to be aware of the potential for different views to be expressed and for alternative categorisations and explanations to be valid. The researcher must be aware of her/his own preconditioning and views − potential bias.

Analysis of data

Hammersley and Atkinson (1983) are amongst a number of authors who consider the construction of *typologies* and *taxonomies*, which are categories and groups within the categories, to be important elements of analyses. The researcher should seek to establish categories, sub-groups and relationships between them from the data collected. Such categorisation of data will reduce the number of potential variables, thereby making the data more manageable and 'visible' to assist the detection of patterns and possible dependencies, also called causalities. Clearly, in such qualitative research, much analysis is carried out by the researcher during the period of collecting data in the field.

Bogdan and Biklen (1982) differentiate between analysis carried out in the field during the period of collecting data and analysis carried out after the data collection has been completed. They assert that the researcher needs to be engaged in preliminary analyses constantly during data collection whilst post-collection analyses concern developing a system of coding the data primarily.

Charmaz (1983) believes that, 'qualitative coding is not the same as quantitative coding.... Quantitative coding requires preconceived, logically deduced codes into which the data are placed. Qualitative coding... means creating categories from interpretation of the data.' The belief implies that qualitative coding is more flexible, as categories are created to suit the data from the data collected, whereas quantitative coding may require data to be force-fitted into the pre-selected categories.

It is likely to be a poor idea to have a 'miscellaneous' category because, unless the data are allocated to the pre-selected codes as far as possible, albeit with effort to make the data fit into the categories, any difficulty in allocating an item of data will result in its entering the miscellaneous category. The consequence may be that 'miscellaneous' is the largest of all the categories, which may render analyses difficult, if not meaningless!

Quantitative approaches

When are quantitative approaches employed?

Essentially, quantitative approaches involve making measurements by collecting data. The approach is built upon previous work which has developed principles, laws and theories to help to decide the data requirements of the particular research project. Two major questions are: what is to be measured; and how should those measurements be made? The answers are derived from examination of the theory and previous research findings together with the aim and objectives of the research to be carried out; in particular, the hypothesised relationships in the research model. Therefore, the coding framework is, as noted above, already in place.

In allocating data to categories in a database by means of coding exercises, Fine (1975) demonstrated some significant facets which are often overlooked. It seems to be a common belief that the more detailed or complex the database is, the more useful it will be. The main assumption is that of accuracy in allocation – that using the coding system, the data are allocated accurately.

> *Example*
> Fine (1975) conducted an experiment in which construction 'cost accountants' were asked to allocate items to a cost database. The results were:
>
> 30 categories in the database, 98% items allocated accurately
> 200 categories in the database, 50% items allocated accurately
> 2000 categories in the database, 2% items allocated accurately.

Of course, the significance of misallocations can vary enormously depending on the nature of the items, the database, the nature of misallocations and the use of the database.

Given the early stages of quantitative methods to arrive at the field-work stage of the research, there is the requirement of a considerable amount of pre-conception in deciding what data are to be collected, how they will be collected and what analyses will be done. Without thorough study of the underpinning theory and literature, important factors are likely to be missed thereby, at the very least, reducing the validity of outputs from the research as well as causing difficulties in executing the work.

Sources of data

Ideally, the researcher and the existence of the research will have no influence on the data collected. However, that is known to be untrue and so the pragmatic objective is to minimise the impacts. Such minimisation is sought by using objective methods designed to remove as much bias as possible and to conduct the research in the most unobtrusive way, whilst retaining goodwill of the collaborators and subjects of study – essential in studies of people and their behaviours.

Throughout quantitative studies, and scientific method, a major objective is that the research is 'value-free'; that the work is unaffected by the beliefs and values of the researcher(s) – it is objective. In conducting quantitative research, three main approaches are employed: asking questions of respondents by questionnaires and interviews; carrying out experiments; and 'desk research' using data collected by others. Using data collected by others, who collected it possibly for a variety of other purposes, can be problematic, as the data, sampling etc. have not been tailored to the particular research project in question. However, it can be very helpful to use data collected already by others – it saves time, can be cheap and, for studies such as macro-economics, can be the only viable way of obtaining the data required.

Clearly, it is essential to investigate the nature of the data and collection mechanisms in order to be aware of the limitations of the data and their validity, notably comparability – for instance, the basis for producing unemployment statistics in the UK has changed many times over the last decade. For longitudinal studies, the difficulties of comparing like with like may be great.

In executing experiments, results are sought by effecting incremental changes in the independent variable and measuring the effect, if any, on the dependent variable, whilst holding all else constant. Using the experimental style in a social context produces problems far in excess of those encountered in a science research laboratory; society is dynamic and the number of variables operating is vast. Hence it is highly unlikely that only one variable will change during the study – the practical approach is to restrict the impact of variables as far as possible.

Usually, the sample of people on whom the experiment is to be conducted are split into two groups which are matched as far as possible in all respects. The independent variable is changed for one group but is unaltered for the other, known as the *control group*. Having matched of the samples in the two groups and holding constant all other variables which, in practical terms, act identically on both groups, it is therefore valid to assume that changing the independent variable will yield identical consequences in respect of the dependent variable. Hence, the differences measured between the two groups in respect of the dependent variable over the course of the experiment are due to the measured changes made to the independent variable. Thus, cause and effect in direction and magnitude are established. (The stages in experimental design are shown in Fig. 1.3.) Clearly, very large samples of people are necessary for experimental techniques to be used in investigating behaviours, safety of medicines etc.

Experimental design in human behaviour involves developing strategies for executing scientific inquiry to enable the researcher to make observations and interpret the results. There are two important aspects to consider in formulating experimental design:

- Unit of analysis
- Time dimension.

It is critically important to identify the unit of analysis, such as the individual or the group, accurately. Failure to do so may result in two errors of logic: the ecological fallacy; and reductionism. The ecological fallacy involves gathering data on one unit of analysis but making assertions regarding another. Reductionism refers to an over-strict limitation on the kinds of concepts and variables to be considered in understanding a social phenomenon. Both errors involve misuse of the unit of analysis.

Another aspect of experimental design is the time dimension, as exemplified in *cross-sectional studies,* which are observations at one point in time, and *longitudinal studies,* observations made at multiple time points. The three types of longitudinal studies are: *trend studies,* those that examine changes within some general population over time; *cohort studies,* those that examine more specific subpopulations as they change over time; and *panel studies,* those that examine the same set of people over time (Babbie 1992). Generally, longitudinal studies are superior to cross-sectional studies for making causal assertions.

In formulating a good experimental design, it is important to consider:

- How to vary an independent variable to assess its effects on the dependent variable.
- How to collect data, or in the case of social behavioural research, how to assign subjects to the various experimental conditions.
- How to control extraneous variables that may influence the dependent variable.

In experimental design, the researcher is interested in the effect of the independent variable on one or more dependent variables. A dependent variable is the response being measured in the study. In a between-subjects design, subjects are randomly assigned to experimental conditions, and data can be collected from various groups of subjects. In within-subjects design, all subjects serve in all experimental conditions.

Example

Leary (1991) gives the example of testing the effect of caffeine, the independent variable, on subjects' memories, the dependent variable. The independent variable can be varied in terms of quantitative differences, such as different quantities of caffeine given to the subject, to test its effect on the dependent variable, in this case, the subjects' memories of a list of words. In other circumstances, the independent variable could be varied in terms of qualitative differences, such as the colour of a cup of coffee. One level of the independent variable can involve the absence of the variable of interest. Subjects who receive a non-zero level of the independent variable compose the experimental groups and those who receive a zero level (say, no caffeine) of the independent variable constitute the control group. Although control groups are useful in many experimental designs, they are not always necessary.

In a between-subjects design, each group is given a different level of caffeine consumption; the researcher must be able to assume that subjects in the various experimental groups did not differ from one another before the experiment began. This ensures that, on average, subjects observed under the various conditions are equivalent. Alternatively, in within-subjects designs, a single group of subjects is given all levels of caffeine consumption. As such, the researcher is testing the differences in behaviour across conditions within a single group of subjects.

Within-subjects design is better than single-subjects design for detecting effects of the independent variable. This is because the subjects in all experimental conditions are identical in every way so that none of the observed differences in responses to the various conditions can be due to pre-existing differences between subjects in the groups. Since repeated measures are taken on every subject, it is easier to detect the effects of the independent variable on each subject.

Experimental control

Experimental control is essential; it refers to eliminating, or holding constant, extraneous factors that might affect the outcome of the study. (See also: validities – Chapter 5; obtaining data – Chapter 6.)

Two important types of variance may occur (Leary 1991):

- *confound variance*, which is a between-group variance
- *error variance*, which is a within-group variance.

Confound variance is that portion of the variance in subjects' scores which is due to extraneous variables that differ systematically between the experimental groups. Confound variance may be eliminated through careful experimental control, in which all factors other than the independent variable are held constant or subjected to controlled variation between the experimental conditions. An experiment has internal validity when it eliminates all potential sources of confound variance. Internal validity is the degree to which a researcher draws accurate conclusions about the effects of the independent variable. Error variance is due to neither the independent variable (since all subjects in a particular condition receive the same level of the independent variable) nor the confounding variables, since all subjects within a group would experience the same confound. Error variance is due to individual differences among subjects within the group, such as ability, personality, mood, past history etc., random variations in the experimental setting and procedure, perhaps the time of testing, differential treatment of the subjects and measurement error.

Nothing other than the independent variable should vary systematically across conditions, otherwise, confounding occurs. Confounding destroys the internal validity of the experiment, which is a very serious flaw. Researchers should also try to minimise sources of error variance which is produced by unsystematic differences between subjects within experimental conditions. Although error variance does not undermine the validity of an experiment, it makes it more difficult to detect the effects of the independent variable. However, attempts to reduce the error variance in an experiment often lower the study's external validity; the degree to which the results can be generalised.

In most instances, researchers do not have the necessary control over the environment to structure the research setting, and *quasi-experimentation* results (Condray 1986). Quasi-experimentation is a pragmatic approach which attempts to collect the most meaningful data under circumstances that are often less than ideal. Rather than adhering to just one particular experimental design, researchers may use whatever procedures are available to devise a reasonable test of the research hypothesis. Most often, given the absence of random assignment of subjects, as in between-subject design, simply showing that a particular quasi-independent variable was associated with changes in the dependent variable may not be convincing enough to show that the quasi-independent variable caused the dependent variable to change. The researcher may also have to demonstrate that the quasi-independent variable was associated with changes in other variables assumed to mediate the change in the dependent variable. By looking at other additional quasi-independent variables, comparison groups and dependent measures, researchers increase their confidence in the inferences they draw about the causal link between the quasi-independent and dependent variables.

Collecting data from respondents

Surveys

Much research in the social sciences and management spheres involves asking and obtaining answers to questions through conducting surveys of people by using questionnaires, interviews and case studies. Often, responses are compared to 'hard data', such as the cost records of a project. However, many 'hard data' are not totally 'objective' or reliable in the sense of showing what they may be believed to show.

> *Example*
> Often project costs, known as the *out-turn*, are influenced by the negotiating powers and skills of the project participants and so, whilst quantifying cost, a factor of human objectives and skills is incorporated in the final cost figures which result.

Given that there is a finite amount of resources available for carrying out the field work, especially where those resources are very restricted, a choice of research method is necessary. The choice is affected by consideration of the scope and depth required. The choice is between a broad but shallow study at one extreme and a narrow and deep study at the other, or an intermediate position – as shown in Fig. 4.2

Survey techniques, such as questionnaires, interviews etc., are highly labour intensive on the part of respondents and particularly

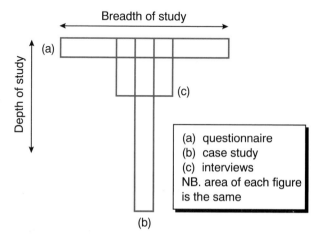

Fig. 4.2 Breadth v. depth in 'question-based' studies.

on the part of the researcher; one consequence is the low response rate, which is common, notably for postal questionnaires which can expect a 25%–35% useable response rate. Thus, many surveys do not produce data from which results capable of strong support for tests of hypotheses or conclusions can be drawn. Further, self-completed responses are very prone to bias and distortions – giving answers which respondents believe 'should' be given rather than providing their 'true' answers; giving answers to 'please' the researcher etc. – as well as being self-perceptions by the respondents.

Questionnaires

Questions occur in two primary forms – open or closed. Open questions are designed to enable the respondent to answer in full; to reply in whatever form, with whatever content and to whatever extent the respondent wishes (in interviews, the researcher may probe). Such questions are easy to ask but may be difficult to answer, the answer may never be full/complete and, often, the answers are very difficult to analyse. It is essential that answers to open questions are recorded in full. Closed questions have a set number of responses as determined by the researcher. However, such rigidity of available responses may constrain the responses artificially, hence a response opportunity of 'other, please state' should be provided wherever possible. Care must be taken that responses to open questions are not biased by the response alternatives provided by related and preceding closed questions. Thus, it may be preferable to place open questions before related, closed questions. It is possible to ask more closed than open questions, as responses to closed questions can be given more easily and quickly.

Filter questions may be employed to progress certain respondents past a set of questions which are not relevant to them. Although the technique speeds respondents through the survey and maintains relevance of the questions answered, extensive use of filter questions can be annoying. Funnelling can be used where issues are introduced by a general question and pursued by questions of increasing detail on aspects of the issue being researched. Especially for attitude studies, a form of funnel-sequence of questions is likely to be helpful:

Example
1. Questions to discover the respondents' awareness/knowledge/ thoughts about the issue.
2. Questions to discover the respondents' general views of the issue (open questions).
3. Questions concerning specific facets of the issue being researched (closed questions).
4. Questions to discover reasons for the views held by the respondents.
5. Questions to discover how strongly the respondents hold their views.

Questionnaires may be administered by post to respondents, to groups by the researcher or particular individuals, such as to a class of students, by a lecturer, or to individuals by the researcher – perhaps to form the basis of an interview. The questions should be unambiguous and easy for the respondent to answer, they should not require extensive data gathering by the respondent. They should not contain unnecessary requests for data, for instance, they should not request a name when the respondent is known, since the question-naire was sent to the person by name, especially when anonymity is to be provided or when the identity of the respondent is not needed. Questions should be clear, each should concern one issue only and the request for answers should be given in an 'unthreatening' form appropriate to the research.

Example
Seek a respondent's age by requesting that they tick the appropriate 10 year age band, ensuring the bands do not overlap – e.g. 10–19, 20–29 etc.

Questions may seek opinions on an issue by degree of agreement with a statement, phrased objectively – it is preferable to use a four point rather than a three or five point scale for the answer (strongly agree, agree, disagree, strongly disagree) to avoid any 'opting out' or indecision by respondents' selecting the middle, non-committing answer.

Questions concerning ordering of hierarchies of issues, derived from literature, should not contain too many items to be ranked; such questions should include the opportunity for respondents to note their own items which do not appear on the list – include an item of 'other, please note'. A useful alternative is to ask respondents to rank their top 5 or 6 items.

If a statement from the respondent in answer to a question is requested, ensure that the statement required can be provided in a few words (the space provided on the form will suggest the size of answer required). Long statements are tedious for respondents and may be difficult to analyse.

Questions concern facts, knowledge and opinion. It is important to appreciate that people's memories are imperfect. Whilst facts may be checked in some cases, and a respondent's knowledge level can be assessed, opinions must be taken at their 'face value' which may be problematic in instances where opinions expressed are inconsistent.

All questionnaires should initially be piloted; completed by a small sample of respondents. The piloting will test whether the questions are intelligible, easy to answer, unambiguous etc., and through obtaining feedback from these respondents, there will be an opportunity for improving the questionnaire, filling in gaps and determining the time required for, and ease of, completing the exercise. Discussion of the questionnaire with the supervisor and other researchers is a useful supplement to the piloting, as it provides a research-oriented view of the questions, the components and assembly of the questionnaire and probable approaches to the analysis of responses.

Interviews

Interviews vary in their nature, they can be:

- structured,
- semi-structured and
- unstructured.

The major differences lie in the constraints placed on the respondent and the interviewer. In a structured interview, the interviewer administers a questionnaire, perhaps by asking the questions and recording the responses, with little scope for probing those responses by asking supplementary questions to obtain more details and to pursue new and interesting aspects. In unstructured interviews, at the extreme, the interviewer introduces the topic briefly and then records the replies of the respondent. This may be almost a monologue with some prompts to ensure completion of the statements; clearly the respondent can say what and as much as she/he desires. Semi-structured interviews fill the spectrum between the two extremes. They vary in form quite widely, from a questionnaire-type with some probing, to a list of topic areas on which the respondent's views are recorded.

The inputs of the interviewer are critical – especially probings – as the questions asked, the *probes*, will influence the responses obtained. The non-verbal communications or 'body language' of the participants will have an impact on the responses and recordings. Often, with permission of the respondents, tape recording the interviews can be very helpful at the later stages of analysis and, through subsequent scrutiny, help to ensure accuracy and objectivity in recording responses. Transcribing is lengthy, tedious and expensive so the tape recordings may be employed to supplement the interviewer's notes.

Case studies

Often, case studies employ a variety of data collection techniques. Unlike questionnaires and interviews when the case researched is the respondent and so a possibly large number of cases are researched for statistical significance, in a case study the case is the particular occurrence of the topic of research. It may be, for instance, a legal case hearing, a building in use over a time, or the procurement of a construction project. Interviews may be used accompanied by collection of 'hard' documentary data. Questionnaires are less usual although they may be employed to gain an understanding of the general situation of which the case being studied is a particular instance. A case study yields deep but narrow results. Commonly, it will employ triangulation both in the case study itself and to facilitate generalisation of findings. However, it is essential to be aware of the validity of generalising the findings of a case study research project.

Triangulation

Triangulation is the use of two or more research methods to investigate the same thing, such as experiment and interviews in a case study project. A postal or other questionnaire to a generalised, representative sample of respondents would assist the researchers to appreciate the general validity of the findings from the particular case study and would serve to aid understanding of its unique and generally applicable features.

Further techniques of data collection involve asking respondents to keep diaries over particular periods and/or for researchers, possibly aided by cameras etc, to make observations, akin to a laboratory notebook in natural science experiments. Whatever methods are adopted, it is important that they are implemented as rigorously as possible to try to avoid bias and to obtain appropriate amounts of accurate data.

It is important to be aware of methodological considerations, the advantages and disadvantages of particular methods, error sources, possible bias, strengths of triangulation etc. in order that the validity of the study and, in particular, its results and conclusions can be appreciated and, perhaps, quantified.

Summary

It is important that empirical work should not begin until the review of theory and literature has been carried out. Problems of placing too much reliance on experience have been considered. The main approaches to collecting data for qualitative studies have been reviewed, notably the use of 'grounded' theory. Similarities between making measurements and collecting quantitative data were noted; problems of data allocation and coding were reviewed. Difficulties in conducting 'experiments' in a social context have been considered – especially for the handling of different types of variables and establishing relationships between them, especially the issue of causality. Issues in collecting data via questionnaires, interviews and case studies have been examined.

References

Babbie, E. (1992) *The Practice of Social Research*, Wadsworth, Belmont, California.

Bogdan, R.C. & Biklen, S.K. (1982) *Qualitative Research for Education*, Allyn and Bacon, Boston, MA.

Charmaz, K. (1983) 'The grounded theory method: an explication and interpretation'. In *Contemporary Field Research*, (ed. R.M. Emerson), Little, Brown, Boston, MA.

Chau, K.W. (1995) Monte Carlo simulation of construction costs using subjective data. *Construction Management and Economics*. **13**(5), 369–383.

Chau, K.W. (1997) Monte Carlo simulation of construction costs using subjective data: response, short communication. *Construction Management and Economics*. **15**(1), 109–115.

Churchman, C.W., Ackoff, R.L. & Arnoff, E.L. (1957) *Introduction to Operations Research*. Wiley, New York.

Clegg, S.R. (1992) Contracts Cause Conflicts. In: *Construction Conflict Management and Resolution*, (eds P. Fenn & R. Gameson), E. & F.N. Spon.

Condray, D.S. (1986) Quasi-experimental analysis: a mixture of methods and judgement. In: *Advances in Quasi-experimental Design and Analysis*, (ed. W.M.K. Trochim), Jossey-Bass, San Francisco.

Fellows R.F. (1996) Monte Carlo simulation of construction costs using subjective data: a note. *Construction Management and Economics*. **14**(5), 457–460.

Fine, B. (1975) 'Tendering strategy'. In: *Aspects of the Economics of Construction*, (ed. D.A. Turin), George Godwin, London.

Glaser, B.G. & Strauss, A.L. (1967) *The Discovery of Grounded Theory: Strategies for Qualitative Research*, Aldine, Chicago.

Hammersley, M. & Atkinson, P. (1983) *Ethnography, Principles and Practice*, Tavistock, London.

Hicks C.R. (1982) *Fundamental Concepts in the Design of Experiments*, 3rd edn, Holt-Saunders International.

Leary, M.R. (1991) *Behavioural Research Methods*, Wadsworth, Belmont, California.

Levin, R.I. & Rubin, D.S. (1991) *Statistics for Management*, 5th edn, Prentice- Hall, Englewood Cliffs, New Jersey.

Meier, R.C., Newell, W. T. & Pazer, H. L. (1969) *Simulation in Business and Economics*. Prentice-Hall, Englewood Cliffs, New Jersey.

Mihram, G.A. (1972) *Simulation. Statistical Foundations and Methodology*, Academic Press, New York.

Mitchell, G.H. (1969) Simulation. *Bulletin of the Institute of Mathematical Applications*, **5**(3), 59–62.

Morgan, B.J.T. (1984) *Elements of Simulation*, Chapman and Hall, London.

Oakley, J. (1994) Thinking through fieldwork. In: *Analyzing Qualitative Data*, (eds A. Bryman & R.G. Burgess), Routledge, London.

Petersen, R.G. (1985) *Design and Analysis of Experiments*. Marcel Dekker Inc., New York.

Rosenblueth, A. & Weiner, N. (1945) The role of models in science. *Philos. Sci.* **12**, 316–321.

Sayre, K.M. & Crosson, F.J. (ed.) (1963) *The Modelling of Mind: Computers and Intelligence*, Simon and Schuster, New York.

Schatzman, L. & Strauss, A.L. (1973) *Field Research: Studies for a Natural Sociology*, Prentice-Hall, Englewood Cliffs, New Jersey.

Strauss, A.L. (1987) *Qualitative Analysis for Social Scientists*, Cambridge University Press, Cambridge,.

Tesch, R. (1991) Software for qualitative researchers, analysis needs and program capabilities. In: *Using Computers in Qualitative Research*, (eds N.G. Fielding & R.M. Lee), Sage, London.

Chapter 5

Hypotheses

The objectives of this chapter are to:

- discuss the **roles of hypotheses**;

- emphasise the necessity for **objective testing of hypotheses**;

- demonstrate the **role of sampling** for testing hypotheses;

- examine **common statistical measures**;

- consider error types in testing **null hypotheses**;

- discuss **validities**.

Roles of hypotheses

Not all research projects will have hypotheses to be tested. For research which is a fundamental exploration of human behaviour in a topic for which theory has not been developed to any significant extent and, hence, where little or no research has been carried out, it is not possible to generate hypotheses in a meaningful way. The appropriate approach in such circumstances is not to force issues artificially, in this case identification of variables and supposing causal relationships between them, but to observe behaviour whilst minimising the effects of such observation on the subjects' behaviours.

However, for a great deal of research, notably, the application of scientific method, or quantitative studies, it is both possible and important to draw on theory and literature (findings of executed research) to formulate hypotheses to be tested. In both producing a proposal for a research project and in executing the research, the hypothesis acts as the focus for the work and one which helps to identify the boundaries also.

As the early stages of research progress, from the preliminary investigations undertaken to help to produce the proposal, with the review of theory and literature, the main hypothesis and sub-hypotheses may be modified as greater knowledge of the topic and main issues involved is gained. On occasions, especially when working in new areas of study, some researchers prefer not to include an hypothesis in the proposal but to develop the hypotheses once the theory and literature have been examined, such that the hypothesis arises out of a good and up-to-date understanding of the topic.

Provided there is sound appreciation of the dynamism of any research project, which is due to the very nature of the research process, the suggestion of an initial hypothesis, which may be subject to revision, has certain desirable elements. It is essential to recognise that as knowledge advances and increases, so beliefs change, and to adhere to initial ideas rigidly excludes advances. For research to yield its full potential, the developmental aspects must be incorporated. It is important that researchers do not feel either forced to include an hypothesis in a proposal or to be restricted by not being allowed to amend an hypothesis once it has been advanced. However, if too much flexibility is allowed over the provision and/or adaptation of hypotheses, the objectives of the work, as manifested in outputs, may suffer, and it may be that the research hypotheses change constantly from the original, with the result that none are ever tested.

Hence, there is a consideration of what is appropriate – not too much forcing but also a degree of limitation, so that whatever approach is adopted as being suitable for the research project, that approach is followed with rigour. What is done, how, when etc. must be capable of being substantiated by the researcher, who should be able to provide valid reasons for the approach – including changes made to the research plan. Do not do things without good reason. Reasons of practicality can be just as valid and important as reasons of theory.

Objective testing of hypotheses

An hypothesis is *a statement*, a conjecture, a hunch, a speculation, an educated guess. It is the main statement of supposition permissible which will be tested rigorously by the research to remove as much of the supposition or uncertainty as possible and replace it with knowledge, this may be certainty – more realistically, risk or probability. Many research projects are dependent upon assumptions, often due to the nature of the theory on which the research is based. Such assumptions are statements of supposition and, as such, must be identified and stated. If possible, investigating their validity should be conducted and, perhaps more especially, the consequences of relaxing those assumptions on the results of the research.

Chambers English Dictionary defines an hypothesis as:

- a supposition
- a proposition assumed for the sake of argument
- a theory to be proved or disproved by reference to facts
- a provisional explanation of anything.

The fourth definition encapsulates the view of the highly qualitative, ethnographic type of research which seeks to observe in order to establish hypotheses; provisional explanations which may be subject to testing for verification.

The KISS principle is an efficiency principle – Keep It Simple, Stupid. It is a good research principle from several views – keep contents straightforward, simple and clear; write plainly so that the language is easy to read and the arguments can be followed readily and have no ambiguities and, perhaps most fundamentally, search for the simplest basic facts, principles, models etc. This is what mathematicians call *elegance*.

The third definition of an hypothesis raises a problem. Commonly, especially new, researchers believe they should prove or disprove an hypothesis and so they set out to do so. Proving, in the absolute sense, will not be possible, nor should that be a goal as such a goal is likely to introduce *bias*. The goal is to *test* the hypothesis as rigorously as possible; that means *objectivity*. In testing an hypothesis, a researcher should seek to provide evidence, through results of the testing, to support or reject the hypothesis at an appropriate level of confidence and obtained, if possible, through statistical testing and, therefore, quantified. Any level of confidence calculated should be specified in the results – that aids appreciation of the conclusions and their validity.

Moser and Kalton (1971) note – 'Surveys thus have their usefulness both in leading to the formulation of hypotheses and, at a more advanced stage, in putting them to the test. Their function in a given research project depends on how much is already known about the subject and the purpose for which the information is required'. They warn against the forced inclusion of hypotheses in surveys of 'factual enquiries' which should be undertaken only after thought has been given to what should be asked. Forcing hypotheses into such studies inappropriately results in only trivial hypotheses and, thereby, hardly avoids the (poorly-informed) criticism of 'factual surveys' that they do not include hypotheses and their testing! So, it is important to recognise when it is appropriate to include hypotheses in the research and when it is not. It is appropriate to include an hypothesis in research, when it is based on theory and previous work and, being so, sets out to test the existence of certain variables and/or any relationship between them. Quantitative studies are the most obvious instances of research projects which have an hypothesis to test.

An hypothesis is inappropriate for a qualitative study which seeks to carry out a fundamental investigation to identify what is occurring – such as to observe behaviour in a highly novel environment or in a new community in which established values, theories etc. may not apply. There are three requirements for an hypothesis:

- It should be *testable* – so that it may be supported or rejected from empirical evidence.
- It should be *positive* – testing the hypothesis concerns what is, *not* what ought to be.
- It should be expressed in clear and simple language – so that it means the same to everyone.

Role of sampling

Usually, testing an hypothesis involves sampling from the population by collecting data and executing analyses. Hence, statistical techniques play a very important role in testing hypotheses. There are many statistical techniques which can be employed to test hypotheses. Although several of the more common statistical techniques are considered in this book, specialist statistical texts should be studied to gain appreciation of the range of techniques available and to ensure that the most appropriate are used. Further guidance may be obtained from the manuals which accompany computer software statistical packages. Figure 5.1 emphasises appropriate data sets to be used for building, testing and, potentially, modifying 'models' and the relationships between variables.

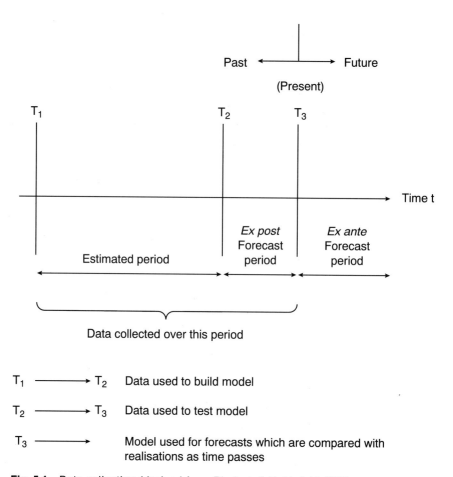

Fig. 5.1   Data collection (derived from Pindyck & Rubinfeld 1981).

As time passes and more data become available, those data may be employed to modify the initial model continuously. However, it is important that data used to build a model are *not* also used to test the model; the testing must be done against data not used in the model-building, otherwise the model would, inevitably, contain an element of being a 'self-fulfilling prophesy' with consequent distortion to the fit, variability and accuracy measurements.

Further, it should be noted that many measurements are really indicators of varying degrees of error (from the 'true' measure). In dynamic analyses, such as for an economy, indicators are classified as:

- *Leading* (long; short) – 'where we think we will be'.
- *Co-incident* – 'where we think we are'.
- *Lagging* (short; long) – 'where we think we have been'.

Sampling is necessary because it is rarely possible to examine an entire population. It would be possible to survey the techniques used to pre-stress pre-cast concrete components manufactured in Hong Kong as the population of manufacturers of such components is small. It would *not* be possible to carry out a full population survey of the type of tea preferred by the Hong Kong population. The main reasons for the impossibility include population changes (births, deaths, migrations), time, cost, tracing people, obtaining the data. Fortunately, to obtain a good representative picture of the population's tea drinking preferences, it would be possible to use a sample of the population, which is much smaller than the total population but sized and structured to be statistically representative. Clearly, the results obtained from such sampling will not be exactly the same as if the whole population had been consulted, but the result will be adequate for the purpose for which the information is required, such as for an importer to know the appropriate quantities of each type of tea to import annually into Hong Kong.

So, the sample needs to be representative of the population in order to produce a result of theoretical and practical value and that the results obtained from the sample approximate as closely as possible to those which would be obtained if it was possible to survey the entire population.

Common statistical measures

A very common measurement is the average; in statistics, there are two important measurements of the average or *mean*:

- the *arithmetic mean*; which most people understand as the average
- the *geometric mean*.

The arithmetic mean is expressed mathematically as:

$$\bar{x} = \frac{1}{n} \sum_{i=1}^{n} x_i \quad \text{or} \quad \bar{x} = \frac{\sum x}{n}$$

where n = no of items

       x_i = an item of value i where the value of i ranges from 1 to n

       $\bar{x}$ = arithmetic mean.

The geometric mean is calculated as:

$$GM = \sqrt[n]{x_1 x_2 \dots x_n}$$

In sampling, the mean of the sample is derived to approximate, as closely as practicable, to the mean of the population. Distributions are very important too – averages can be very misleading.

Example
A statistician once illustrated the point: 'if I have one foot in a bucket of ice and the other foot in a bucket of boiling water, on average, I'm OK!'

Is s/he?

Clearly the statistician would not be 'OK', so it is important to consider measurements of dispersion, distribution or variability too. The *median* is the mid point of a distribution. As 'outlier' values may distort the mean, the median may be a preferable indicator of such a distribution's central tendency. The *mode* is the value which occurs most frequently in a distribution. A distribution may have more than one mode.

Example

(a)   Symmetrical distribution (e.g. normal distribution):

mean, mode and median coincide.

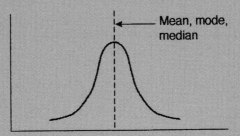

(b)   Skewed distribution

Mean and, to a lesser extent, median move away from the mode in the direction of the skew (long tail).

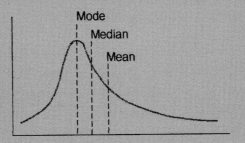

Note: 'outlier' points of the distribution affect the location of the mean more than the median and the median more than the mode. Hence, if 'outliers' cannot be disregarded, their effects must be acknowledged.

(c)   Bimodal distribution

Note: The modes need not be of equal size or value.

The arithmetic mean is the value which, numerically, is the most representative of the series, or distribution. The geometric mean is used where measurements of relative changes are being considered.

Example

The marks gained by a group of students form a frequency distribution (frequencies being the number of students obtaining each mark); for a frequency distribution, the arithmetic mean is:

$$\bar{x} = \frac{\sum fx}{\sum f}$$

where $\sum f$ is the sum of the number of students sampled obtaining each mark.

The geometric mean is used in retail price indices; for such variables, it is usual to be interested in the percentage change between successive periods.

Samples, and the populations from which samples are drawn and which the samples should represent, each have statistical distributions – commonly assumed to be normal; tests for normality can be applied.

However, many populations are structured in a particular way which is important to the research. In order to obtain a sample which is representative of the population, the sample should reflect the structure of the population closely. The structure of the population yields the sampling frame – the structure used to determine the form of the sample from which the data will be gathered.

Generally, sample surveys have the objective of either:

● estimating a parameter in the population – the estimate made should be accompanied by a note of its precision
● testing a statistical hypothesis concerning the population – the results of testing will lend support to or lead to rejection of the hypothesis; such tests require a criterion against which deviation of the result from the sample can be judged against the value hypothesised.

In both instances, the measure of precision used is the standard error.

It is important to note that statements which are made on the basis of results derived from samples are probability statements and that random sampling should be employed. Random sampling is where each member of a population has a known and non-zero probability of being included in the sample. If a sample of size n is taken an infinite number of times from a population by a random sampling method, the distribution of the sample means is the sampling distribution of the mean.

An *estimator*, the method of estimating the population parameter or measurement from the sampled data, should be unbiased; it is unbiased if, on average over an infinite number of samples, the *sample estimates*, (measurements yielded by the estimator, such as the sample mean, $\bar{x}$) equal the *population parameter* (in this case, the population mean, μ).

The *standard error* of the means measures the degree to which estimates which are obtained from different samples are likely to differ from each other.

$$\text{Standard error } (\bar{x}) = \sqrt{\frac{\sigma^2}{n} \times \frac{N-n}{N-1}}$$

where σ = standard deviation of the population parameter
 N = size of the population
 n = sample size

As illustrated in Fig. 5.2, the inaccuracy of an estimate obtained from a sample is measured by the extent to which it differs from the population parameter; an estimator is unbiased if the mean of the estimates obtained from all possible samples is equal to that of the population parameter. The standard error is a measure of the fluctuations of estimates around their own mean (shown by the sampling fluctuations). Hence, the reference point of the standard error is the expected value rather than the population value, as it is a 'sample-based' measure.

The *mean square error* (MSE) measures the variability around the population value:

$$\text{MSE} = Var(\bar{x}) + (\mu - m)^2$$

where: $Var(\bar{x})$ is the variability around the expected value
 μ is the population value
 m is expected value of the estimator; the mean of the estimates obtained from all possible samples
 $(\mu - m)$ is the bias of the sample.

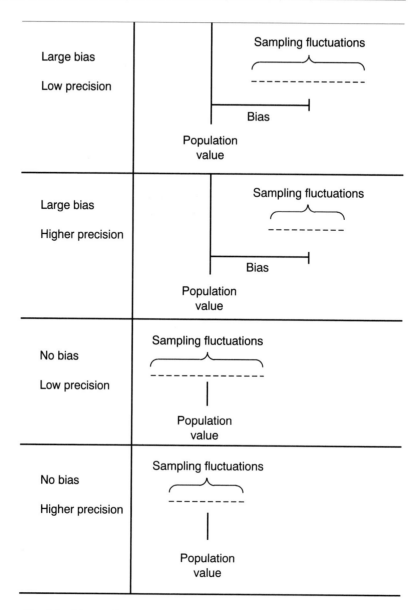

Fig. 5.2 Bias and precision (source: Moser & Kalton 1971, following Deming 1950).

Normal distribution

Much of statistics uses a particular distribution which has been found to occur very commonly – the *normal distribution*. The normal distribution is a probability density function which has certain, particular properties.

Figure 5.3 shows the normal distribution. The main features of the normal distribution are:

- measurements of the random variable, X, are on an interval scale or a ratio scale,
- values of X are symmetrical about the mean, μ,
- the distribution has a central tendency; i.e. the closer values of X are to μ, the more frequently they occur,
- the tails of the distribution spread to infinity.

Thus, the distribution is 'bell-shaped'.

Another important measurement for a normal distribution concerns the variability or spread of the distribution. That measurement is the *variance*; the square root of the variance is the standard deviation, σ.

As, in practice, samples are used, *the standard deviation* of the population must be estimated from that of the sample:

$$\text{Standard deviation of samples} = \sqrt{\frac{\sum (x_i - \bar{x})^2}{n - 1}}$$

For samples of $n \geq 32$, it is acceptable to use n rather than $n - 1$.

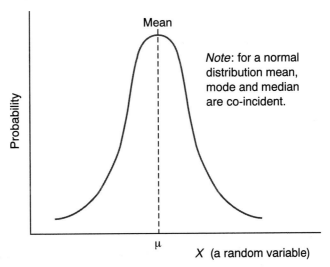

Fig. 5.3 Normal distribution (probability density function).

In many instances, the standard deviation is employed to measure the dispersion as it is on the same scale of measurement as the mean; hence the standard deviation is a more convenient measurement than the variance.

Because the normal distribution is described by a particular formula, from knowledge of the mean and standard deviation, the areas under the curve are known too:

the range bounded by the curve between $x = \mu \pm \sigma$ contains approximately 68% of the area under the curve,
the range bounded by the curve between $x = \mu \pm 2\sigma$ contains approximately 95.5% of the area under the curve,
the range bounded by the curve between $x = \mu \pm 3\sigma$ contains approximately 99.7% of the area under the curve.

Measurements of the spread of a distribution which are simpler than the standard deviation are:

- *range* – susceptible to distortion by extreme values ('outliers')
- *interquartile range* – the range between the first and third quartiles
- *decile range* – the range of the distribution excluding the lowest and highest 10% of data points (i.e. 80% range of the distribution).

The interquartile and decile ranges use the median as the centre of the distribution.

Given a particular value from a sample, the level of confidence concerning its variability can be calculated *provided* it can be established that it is reasonable to say that the normal distribution is applicable and therefore parametric tests are appropriate.

To calculate the level of confidence, the *standard normal* variable, z is employed:

$$z = \frac{\bar{x} - \mu}{\sigma}$$

The usual levels of confidence are:

- 5% is described as significant
- 1% is described as highly significant
- 0.1% is described as very highly significant.

Null hypotheses

Another form of hypothesis which is encountered in statistical testing is the *null hypothesis*. Commonly, null hypotheses are employed in supplementing the overall hypothesis and any sub-hypotheses and they lend rigour to statistically testing particular, possible relationships between variables. A null hypothesis is tested in comparison with its opposite, the alternative hypothesis. The usual form is that the null hypothesis speculates that there is no difference between, say, the cost in $ per kilometre of travel by coach and by train. The alternative hypothesis speculates that there is a difference. Thus:

$$H_O: \mu_1 = \mu_2$$
$$H_A: \mu_1 \neq \mu_2$$

where: $\mu_1 = \text{population mean}_1$
$\mu_2 = \text{population mean}_2$
$H_O = \text{null hypothesis}$
$H_A = \text{alternative hypothesis}$

Similarly, the approach can be used to examine the value of the mean of a sample and that of the population.

$$H_O: \mu = \$\alpha/\text{km}$$
$$H_A: \mu = \$\alpha/\text{km}$$
$$\alpha = \text{constant}$$

To test H_O and H_A, the *t*-distribution is used. The calculation of *t* will be compared with the value obtained from the table of the *t*-distribution.

$$t(\text{calc}) = \frac{\bar{x} - \mu}{S/\sqrt{n}}$$

where: $\bar{x} = \text{sample mean}$
$\mu = \text{population mean}$
$S = \text{sample standard deviation}$
$n = \text{number in sample}$

The level of confidence (or significance) must be decided in order to determine the appropriate value of *t* from the table of the *t*-distribution (*t*(tab)). The degrees of freedom must be calculated to select the correct t(tab). For a t-distribution, there are $(n-1)$ degrees of freedom to determine t(tab). So, if t(tab) $> t$(calc), the null hypothesis is accepted (i.e. $\mu = \$a$/km at the selected level of confidence). However, this result does not substantiate that the null hypothesis is true, it shows that there is no statistical evidence to reject it.

To consider $H_O: \mu_1 = \mu_2$ against $H_A: \mu_1 \neq \mu_2$, again the appropriate level of confidence/significance must be decided (commonly 5%). Essentially, the test concerns whether the means of the samples, $\bar{x}_1$ and $\bar{x}_2$ are drawn from populations with the same mean. The means of the populations are unknown. The method of testing requires the sampling distribution of the statistic $(\bar{x}_1 - \bar{x}_2)$ to be constructed, this is a normal distribution (Yeomans, 1968), the mean of which is zero when $\mu_1 = \mu_2$. The method requires all possible means of each sample distribution to be calculated. All possible pairs of the sample means are considered, one mean being subtracted from the other (i.e. $\bar{x}_1 - \bar{x}_2$). Then, if the mean of the resultant sampling distribution $(\bar{x}_1 - \bar{x}_2)$ is zero, H_O is accepted.

A particular feature of null hypotheses is whether a 'one-tailed' or a 'two-tailed' test is required. If the null hypothesis is of the form $H_O: \mu_1 = \mu_2$, a two-tailed test is necessary where the alternative hypothesis considers $\neq$, in which both greater than or less than values are considered. However, if the alternative hypothesis is of the form $H_1: \mu_1 < \mu_2$, a one-tailed, (left-tailed, lower tailed) test is required (if $H_1: \mu_1 > \mu_2$, a one-tailed, right-tailed, upper-tailed test is required). For a two-tailed test, the region for rejection of the null hypothesis is in both tails of the distribution whilst in the one-tailed test, the region for rejection lies in the appropriate of either the lower or upper tails.

Sometimes testing an hypothesis yields an incorrect result – such error may occur in two ways. A type 1 error occurs when a null hypothesis is correct but the decision taken, based on a test, is to reject H_O, i.e. H_O is rejected incorrectly. A type 2 error is when H_A is correct (H_O is, therefore, incorrect) but the result of testing leads to the decision to be to reject H_A, and accept H_O, incorrectly. Conventionally, the risk of making a type 1 error is α and the risk of making a type 2 error is β. Ideally $\alpha = \beta = 0$ but, as most data are obtained through sampling (rather than measuring the entire population), the practical objective is to minimise α and β and to keep them small.

The significance level chosen for testing an hypothesis is α, the probability of making a type 1 error. Depending on the consequences of making a wrong decision (from consideration of the null and alternative hypotheses), a decision maker will be able to say which type of error is more onerous – i.e. which type of error is worse; if it is a type 1 error (falsely rejecting a null hypothesis which is correct) a high level of significance (α at 1% or 0.1%) should be chosen. The question which remains is, 'what is the size of β'?

The value of $(1 - \beta)$ measures the probability of rejecting a null hypothesis when it is false; this is the objective of testing null hypotheses. Hence, high values of $(1 - \beta)$ are desirable. $(1 - \beta)$ is called the power of the test so, the smaller the value of β, the greater is $(1 - \beta)$ and, therefore, the power of the test. As changes in the sizes of α and β work inversely (as a trade-off) if α is reduced, β will increase.

Validities

Frequently, research is concerned with investigating an hypothesised causal relationship between an independent variable and a dependent variable; if such a relationship is found, inferences are drawn about the population and, perhaps, a variety of circumstances in which the relationship may apply beyond those of the particular study carried out. Such research involves a set of *validities* (the likely truth of a hypothesis). They are described as construct, internal, statistical inference and external validities. (See also: experimental control – Chapter 4; obtaining data – Chapter 6.)

Construct validity concerns the degree to which the variables, as measured by the research, reflects the hypothesised construct. Poor construct validity occurs if the measurements are caused by other variables' influence or random noise.

Internal validity is high where the observed and measured effect is due to the identified causal relationship. To achieve good internal validity, care is needed in the research design such that alternative explanations are examined and appropriate methods selected by which the causality can be investigated.

Statistical inference validity, judged by inference statistical measurements, is high where the sample is a good representation of the population. Hence, effects on the population can be inferred with a high level of confidence from the behaviours of the sample because the statistics of the sample are close approximations to the parameters of the population.

External validity concerns the degree to which the findings can be generalised over circumstances which are different from those of the tests carried out. It concerns the questions of how restrictive is the study; are the findings applicable to other populations? Judgement of external validity requires careful comparison of the sample and the population from which it was drawn with other populations. This should include a comparison of circumstances for the two populations. For in-depth discussion of validities, refer to Dooley, 1990.

Jung (1991) discusses a possible inverse relationship between the internal and external validities of experiments – the 'experimenter's dilemma' – high internal validity tends to produce low external validity and vice-versa. Commonly, external validity is not a crucial consideration, although it may be important. (Mook, 1983); the results of a particular study often depend too greatly on the context to facilitate generalisation of the findings, but internal validity is essential. In some cases, external validity is important, for instance, surveys which aim to predict from a sample to the population. Frequently, in experimental research, the purpose is to test hypotheses about the effects of certain variables – experiments are designed to determine whether, and to what extent, the hypotheses are supported by the data. If the evidence supports the hypotheses, and hence the theory, the results may be used to generalise the theory, not the results themselves, in other contexts. Generalisability of a theory is determined through replicating experiments in different contexts, with different subjects and, further, by using modified experimental techniques and procedures.

Cordaro and Ison (1963) consider the possible effects of the experimenter expecting particular results, and the resultant consequences. Usually, researchers have an idea or expectation of how subjects will respond to questions, or how materials in laboratory experiments will behave etc. – frequently such expectation, or its opposite, provides the explicit hypothesis for the research. Thus, it is important to be aware of potential bias in that such 'expectations' can distort the results by affecting how the researchers interpret data such as responses etc. To help to counteract any such bias, the researcher might adopt an hypothesis which is the opposite to the researcher's expectations.

Summary

This chapter has considered the definition of an hypothesis and proceeded to examine how appropriate it is to formulate hypotheses for different types of studies. Where appropriate, hypotheses aid both objectivity and delineation of the parameters of the research. The role of sampling was discussed and basic statistical measures for means and dispersion were noted. The use of null hypotheses was examined and how such hypotheses assist determination of the level of confidence in research findings. Common types of errors in null hypotheses testing were discussed.

References

Cordaro, L. and Ison, J.R. (1963) Psychology of the scientist: X, Observer bias in classical conditioning of the planaria, *Psychological Reports*, **13**, 787–789.

Deming, W.E. (1950) *Some Theory of Sampling*, Wiley, London.

Dooley, D. (1990) *Social Research Methods*, 2nd edn, Prentice Hall, Englewood Cliffs, New Jersey.

Jung, J. (1991) *The Experimenter's Dilemma*, Harper and Row, New York.

Mook, D.G. (1983) In defense of external invalidity, *American Psychologist*, **38**, 379–387.

Moser, C.A. & Kalton, G. (1971) *Survey Methods in Social Investigation*, 2nd edn, Dartmouth, Aldershot.

Pindyck, R.S. & Rubinfeld, D.L. (1981) *Econometric Models and Economic Forecasts*, 2nd edn, p. 204. McGraw-Hill, Singapore.

Yeomans, K.A. (1968) *Statistics for the Social Scientist 2: Applied Statistics*, Penguin, Harmondsworth.

Chapter 6

Data Collection

> The objectives of this chapter are to:
>
> - consider how **data requirements** may be determined;
> - identify issues of **sampling** and **sample size**;
> - discuss different **types of data**, appropriate scales of measurement and tests;
> - examine problems of **obtaining data** from respondents.

Data requirements

At an early stage of a research project, it is a good discipline to give preliminary consideration to data requirements. For any study which extends beyond a review of literature and theory, a major issue is the collection of data. However, just because a researcher wishes to collect certain data does not ensure that those data will be available. Restrictions on collection of data apply for a variety of reasons – confidentiality, ease of collection or provision, cost, time etc.

Despite the potential problems, it is helpful to determine what data are ideally required for the research, and then to modify those requirements, if necessary, to overcome practical difficulties. The objective is to obtain an appropriate set of data which will permit the research to proceed, given the dynamism of research and the practical considerations, with outputs reasonably close to the original intentions.

Essentially, a research project is a form of information system. Figure 6.1 is a simple model of an information system. For a discussion of information systems in construction, see, for example, Newcombe *et al.* (1990) pp. 115–135.

To determine the inputs for an information system, the outputs required must be decided first. After consideration of the conversion process, deciding on the analyses etc. which will be carried out to yield the outputs, the input requirements can be determined. Note that the system is analysed from outputs to inputs. For a research project, start with what outputs are desired, then consider what analyses can be carried out and what alternatives are appropriate. From this one can decide the data (input) requirements. However, the very nature of research, as a 'voyage of discovery', dictates that the outputs will not be known until the research has been done, and may not be known even then.

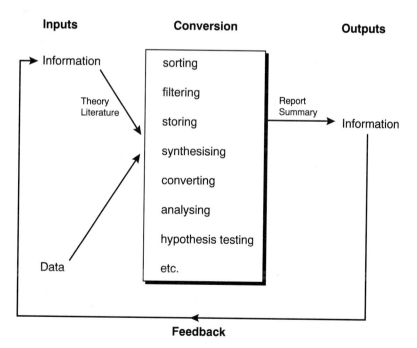

Fig. 6.1 Simple model of an information system.

The research outputs should be considered in terms of the aim, objectives and hypothesis, where relevant, plus sub-hypotheses. So, whilst the outputs in terms of results, findings and conclusions are unknown, the issues which they concern have been determined to a large extent. Such information is available irrespective of the nature of the research project; whether it be a qualitative study involving fundamental investigation of, say, behaviour in a novel environment, where an aim and some objectives will have been established, or a quantitative study, with strict adherence to scientific method, in which aim, objectives and hypothesis will have been formulated at an early stage.

In systems theory, *bounded* systems tend to be 'closed' and, as such, they are isolated from their environment by an impenetrable boundary. In this sense, quantitative studies are bounded, often rigidly, as the variables have been identified and the primary task is to test the hypothesis which has been formulated at the outset of the study or after the review of theory and literature but which remains a precursor to the empirical work. 'Open' systems are *unbounded*, they have a highly permeable boundary, and so the system acts with, and in response to, changes in the environment. Qualitative studies are relatively unbounded as, whilst pursuing investigation of aim and objectives, they collect all possible data to detect variables and relationships etc. Boundaries are important both in their location and their nature – see Fig. 6.2(a) and Fig. 6.2(b). A good discussion of the systems theory approach is contained in Cleland and King (1983).

Sampling

The objective of sampling is to provide a practical means of enabling the data collection and processing components of research to be carried out whilst ensuring that the sample provides a good representation of the population; i.e. the sample is representative. Unfortunately, without a survey of the population, the representativeness of any sample is uncertain, but statistical theory can be used to indicate the representativeness. Measurements of characteristics, such as the mean, of a sample are called statistics whilst those of a population are called parameters. How to obtain representativeness begins with

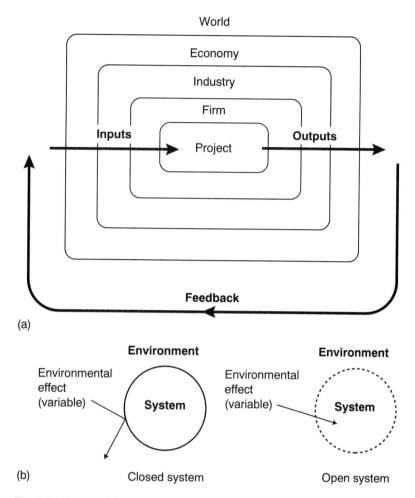

Fig. 6.2 Systems' boundaries.

consideration of the population. Almost invariably, it is necessary to obtain data from only a part of the total population with which the research project is concerned; that part of the population is the *sample*. All buildings on Hong Kong island or all buildings in Greater London can be viewed as populations, whilst both of these are also samples of all buildings in the world. So, the context of the research indicates the population of concern.

The first task is to define the population. If it is sufficiently small, a full population 'sample' may be researched, but in the vast majority of cases a sample must be taken; the question is how? At this point, it is helpful to recall some standard symbols (or notation):

Population size → N;	Sample size → n
Population mean → μ;	Sample mean → $\bar{x}$
Population standard deviation → σ;	Sample standard deviation → s

Given an extremely large sample (e.g. the population of China), whilst the population is actually finite, for practical considerations of time, cost etc., the population is approaching infinite. If there is no evidence of variations in the population's structure, or if there is reason to ignore the structure, *random sampling* will be appropriate. Alternatively, it may be necessary to constrain the sampling – perhaps by considering only one section of the population or/and by reflecting the structure, or a particular structure of that constrained, sampled population. Such 'structured' sampling requires a 'sampling frame' to be established *explicitly*. Within the sampling frame, random sampling, judgemental sampling or *non-random sampling* may be used.

In random sampling, each member of the population has an equal chance of being selected. Usually, such sampling occurs without replacement of the selected members of the population; the members of the population selected are excluded from further re-selection so that each member of the population can be selected once only. The selection of members to be sampled is carried out using random numbers, either from tables or from computer programs; an appropriate way of allocating a unique number to each member of the population must be devised for this sampling technique to be used.

In some instances, judgemental sampling may be used; the judgement, which one hopes will be well-informed, of a person is used to determine which items of the population should form the sample. Such an approach may be very useful in pilot surveys. Alternatively, it is common for judgement to be used to determine which sections, strata or clusters of the population will be sampled; clearly, such sampling may introduce bias, which must be recognised and noted clearly – for example, investigating behaviours of large companies.

Non-random samples are obtained by:

- systematic sampling
- stratified sampling
- cluster sampling.

Systematic sampling does have an element of randomness. Having determined the sample size, every xth member of the population is sampled, where x is the interval between them and is kept constant. Beginning with the yth member of the population, y is selected at random and counted from the first member of the population or some other, appropriate starting point − e.g. people walking past a certain location, beginning at a particular time of day. Given a population of known size, the sampling fraction (n/N) can be used to determine the members of the population to be sampled systematically, where N is the total population and n is the number in the sample. The sampling fraction indicates the members of 'groups' in the population (this will be x). Random numbers provide the starting member of the first group (y), so the members of the samples are y, $y + x$, $y + 2x$, ... The sampling is in the form of an arithmetic progression (or arithmetic series) which sums to $y + (x^2 − 1)$.

Stratified sampling is appropriate where the population occurs in 'distinct', groups or strata. Strata may be selected by others for their own purposes − such as national, published statistics, e.g. the UK statistics of sizes of firms − measured by strata, numbers of employees etc. Strata may be selected for the purposes of the research by, for example, type of firm − industrial, commercial, etc. Ideally, the differences within each group or stratum should be small relative to differences between strata from the perspective of the research project. Having determined the strata sampling occurs, most commonly by considering the relative importance of each stratum in the population and using such weighting to divide n, the sample size, between the strata; the members to be sampled are selected randomly from each stratum.

Cluster sampling is appropriate where a population is divided into groups such that it is likely that inter-group differences are believed to be small, whilst intra-group differences are believed to be large. The population is divided into the groups, called *clusters*, the clusters are selected randomly and the total members of the clusters provide the total sample.

Example

A supplier of steel reinforcement bars is concerned about the accuracy of the cutting to length of straight bars leaving the factory; the concern relates to all diameters of bars supplied; 6 mm, 8 mm,

10 mm, 12 mm, 16 mm, 20 mm, 25 mm, 32 mm, 40 mm. Depending on the practicalities of the situation, stratified sampling or cluster sampling could be used to examine the accuracy of the lengths of bars.

A stratified sample would examine each stratum of bar by diameter, perhaps in proportion to daily factory throughput, by weight, length or cost. A random sample of bars in each stratum would be measured for accuracy of length. A cluster sample would examine, say, a pre-determined number of loads of reinforcement bars, as batched for delivery. Each load would be a cluster, the loads would be selected randomly and would probably each contain bars of a variety of diameters; each load is likely to be somewhat different in the mix of bars it contains. In measuring the lengths of bars in each sampled load, it would be appropriate to note the diameters of the bars measured also.

Sample size

A particular issue in sampling is determination of the size of the sample. By sampling, a statistic called an estimator is obtained. Estimators should predict the behaviour of the population as well as possible – this is achieved by requiring estimators to have four main properties; they should be:

- Consistent
- Unbiased
- Efficient
- Sufficient.

The variance of a consistent estimator decreases as the sample size increases. The mean of an unbiased estimator approximates to the mean of the population; there is an equal chance of the mean of the estimator being more than or less than the mean of the population. This is described by saying that there is no *systematic error*. Systematic error (often called 'bias') is when the errors assume a regular pattern of under- or over-measurement, by a proportion or an amount; such error should be revealed by checking and can be compensated by an adjustment. Further, analysis of the errors may reveal their sources as well as size and that it may be possible to reduce/eliminate the error. Systematic error should be avoidable. Unsystematic error, or *random error*, is almost inevitable, but its size should be kept to a practical minimum by research design and rigour of execution and checking. An efficient estimator has the minimum variance of all possible estimators drawn from the sample. A sufficient estimator is the one which makes most use of the sample data to estimate the population parameter; in particular, the *mean square error* is minimised. Mean square error is an error measure which is used widely; it is the arithmetic average (mean) of the errors squared. Each error is multiplied by itself, which eliminates the sign of the error terms since the square of a negative number is positive. The result is a 'magnified' measure of the average error of the measurements made. The value of the mean square error of the sample is that it provides an unbiased estimate of the variance of the population (σ^2).

Sample sizes are determined by the confidence level required of the estimator. The unknown mean of a population can be estimated with a pre-determined level of confidence as shown below:

- in practical sampling, the sampling distribution of means is a normal distribution; for a large sample $\bar{x} = \mu$, and
- the size of the sample and its standard deviation, s, can be used to estimate the standard error of the distribution, $\sigma_{\bar{x}}$ (Yeomans, 1968).

As s is the best estimate of σ:

$$\sigma_{\bar{x}} = \frac{s}{\sqrt{n}} \left[\text{approximates to } \frac{\sigma}{\sqrt{n}} \right]$$

For a normal distribution, the 95% confidence intervals are at $\mu - 1.96\sigma$ and $\mu + 1.96\sigma$. So, if z denotes the confidence level required, the confidence interval is:

$$\bar{\mu} = \bar{x} \pm z \frac{s}{\sqrt{n}}$$

where $\bar{\mu}$ is the upper and lower confidence limits of the estimate of μ. Normally, 95% or 99% confidence levels are used i.e. $z = 1.96$ or 2.58. So, manipulation of the formula for μ yields the sample size required:

$$n = \frac{z^2 \times s^2}{(\bar{\mu} - \bar{x})^2}$$

As $\bar{x} - \bar{\mu}$ is half of the width of the confidence interval required, neither of their individual values are required. However, the degree of precision of the estimate which is acceptable must be decided.

Example
Consider the prices in dollars of a particular type of car. 'Experience' may indicate that a precision of $\pm 5\%$ is appropriate, which for a car costing about $5000 translates into about $\pm\$250$ (i.e. $\mu - x = \$250$).

If a sample of cars in the UK indicates that $\bar{x}_{UK} = \$5000$ and $s_{UK} = \$700$, what sample size of similar cars in Australia should be employed if 95% confidence is required and an estimate provision of $\pm\$250$ is acceptable?

$$n = \frac{1.96^2 \times 700^2}{250^2}$$

$$= \frac{2.8416 \times 490\,000}{57\,500} = 24.2$$

So, a sample of 25 cars in Australia would yield an adequate solution. Note the assumption that the standard deviation of the sample is the same in each country (i.e. $S_{UK} = S_A$).

An alternative approach to the sampling problem is to determine the probability that the mean of a sample is within a prescribed range of the known mean of the population. So, if the mean price of cars in UK is £9000 with $\sigma = £1200$, what is the probability that the mean price of a sample of 50 cars will lie between £8700 and £9050?

$$\sigma_{\bar{x}} = \frac{\sigma}{\sqrt{n}} = \frac{1200}{\sqrt{50}}$$

$$= £169.71$$

Using z-values to calculate the probability, and given that:

$$z = \frac{\bar{x} - \mu}{\sigma_{\bar{x}}}$$

For the extreme values of $\bar{x}$:

for $\bar{x} = £8700.00$ for $\bar{x} = £9050.00$

$$z = \frac{8700 - 9000}{169.71}$$ $$z = \frac{9050 - 9000}{169.71}$$

$$= -1.79$$ $$= 0.30$$

Areas from z table are 0.4633 and 0.1179. By adding the areas, the probability that the mean of the sample of 50 cars lies between £8700.00 and £9050.00 is 0.5812.

The *central limit theorem* considers the relationship between the shape of the population distribution and the shape of the sampling distribution of the mean. For non-normal distributions of a population, the sampling distribution of the mean approaches normality rapidly as the size of the sample increases, so the mean of the sampling distribution ($\mu_{\bar{x}}$) equals the mean of the population (μ). The central limit theorem, therefore, allows inferences about population parameters to be made from sample statistics validly, without knowledge of the nature of the frequency distribution of the population. Although there are instances where the central limit theorem does not apply, these instances are quite uncommon.

Types of data

Often, types of data are identified in terms of the nature of the scales of measurement used. The essential issue concerning scales is that of:

- uniformity of measurement – of distances between measures, grades or graduations, of the commonality of understanding of the measures
- consistency of measurements.

Nominal or *categorical* scales classify members of the sample, the *responses*, into two or more groups without any implication of distance between the groups. Hence, nominal scales provide a mechanism of classification for responses and respondents.

Ordinal scales are used to rank responses. Again, there is no indication of distance between scaled points or commonality of scale perceptions by respondents. In essence, it provides a hierarchical ordering.

Interval or *cardinal* scales employ equal units of measurement for the scale. Interval scales indicate the order of responses and distances between them. Use of an interval scale permits statements about distances between responses to be made, but not about relationships in ratio terms between scores; this occurs as the zero point for the scale is selected as a matter of convenience rather than having some basic, absolute fixity of reference.

Ratio scales, however, do have a fixed zero reference point, plus the properties of an interval scale. Hence, conclusions can be drawn about both differences in scores (because intervals on the scale are equal – e.g. 1–3, 3–5, 5–7, 7–9), and the relationship between the score – e.g. 8 = twice 4. For a discussion of scales of measurement for variables, see, for example, Bryman and Cramer (1994).

Perhaps the most common scale for obtaining respondents' opinions is the Likert scale. As noted by Bell (1993), such scales are concerned with determining respondents' degrees of agreement or disagreement with a statement on, usually, a 5-point or 7-point scale. By using an odd number of response points, respondents may be tempted to 'opt out' of answering by selecting the mid point. Hence, it may be helpful not only to keep the number of response points small but also to use an even number of response points, thereby having no central point. So, a 4 or 6-point scale of responses may be preferable to the more usual 5 or 7 points.

The next issue concerns what to do with such a scale of responses; the appreciation of the nature of the scale and hence, what can be determined. As a Likert scale is an ordinal scale, it can be used to produce hierarchies of preferences which then can be compared across groups of respondents as per the sampling frame. Using such an approach, it is possible to determine various groups of respondents' views of an issue by asking respondents from each group to respond to a common set of statements against the Likert scale.

> *Example*
> Typical set of statements from which respondents may be asked to choose.
>
> | Either: | (1) | (2) | (3) | (4) |
> |---|---|---|---|---|
> | **A** | Essential | Useful requirement | Helpful addition | Irrelevant |
>
> or
>
> | **B** | Strongly agree | Agree | Disagree | Strongly disagree |
> |---|---|---|---|---|
>
> For **B**, a 5-point scale would employ:
>
> | (1) | (2) | (3) | (4) | (5) |
> |---|---|---|---|---|
> | Strongly agree | Agree | Unsure/Uncertain | Disagree | Strongly disagree |

Analysis of data from such scales is considered in the next chapter.

Osgood *et al.* (1957) developed a technique using semantic differentials. The technique is suitable for examining the meaning of concepts. Normally, a seven point scale is used, with only the extremes' being described by adjectives; the adjectives are polar opposites of each other and different adjectives are used for each dimension to be measured. Positions on the dimensions marked by respondents are scored 1 to 7. To avoid response set bias, it is advisable to vary the positions of the adjectives at the poles (e.g., not all 'good' on the right). The approach requires the researcher to determine the issues to be investigated, the attributes or concepts which are important and suitable adjectives to describe the extremes of each concept's dimension.

The analytic technique for the scores (Moser and Kalton, 1971) is to compare means of the respondents' scores for each subject researched, such that respondents' profiles of each subject in respect of the attributes judged to be important are obtained. Factor or cluster analysis would yield more detail. As respondents are presented with scales, the poles of which are identified and described, they note their response on each attribute dimension relative to the poles.

Hence, provided it is safe to assume that the understanding of the poles is common amongst the respondents, and that they understand that their position on a continuum between those poles is required, it is appropriate to employ mean scores in analysis.

Example
Fiedler (1967) used semantic differentials in his research into leadership – he employed a questionnaire of semantic differentials to produce a Least Preferred Co-worker (LPC) scale. Scoring, using demarcation levels and means, was used to assist examination of managerial styles. (See, for example, Rowlinson *et al.*, 1993). This method has been replicated, with criticisms.

For measures such as described in the above example, the use of means exclusively, with no measures of variability such as variance or standard deviation, may be restrictive to the extent of unreasonableness, thereby rendering the results of dubious use and validity. Median, modal or an index measure might be more appropriate – see, for example, Ahmad and Minkarah(1988), Shash (1993) for indexes and scales with Likert scale data.

Obtaining data

Given the increasing number of research projects, collecting data is becoming progressively more difficult. The people who are targeted as respondents often receive many requests for data and so, as their time is precious, become unwilling or unable to provide data. A good principle is to present the request for data neatly and politely, ensuring that the data can be provided easily, that they are not too sensitive, that the study is of interest to the respondent and that the respondent will get a return commensurate with the effort expended to provide the data. At least, a summary of the research report should be offered and then *provided*.

Having identified the sample, often by organisations, the next step is to identify the most appropriate respondent in each organisation. For a study of quality, the job title to look for could be 'quality manager', 'director of quality', 'quality controller'. The real issue is

to determine which person is at the appropriate level in the organisation to be able to provide the data required for the research. An
initial telephone call will be useful to determine who, if anyone, is
the appropriate person – preferably by name as well as job title. If
the person identified can be contacted by telephone, the study can
be explained *briefly* and, it is hoped, their agreement to provide data
obtained. The respondent should be advised of the nature and extent
of data required, including the time required for completion of the
questionnaire or interview. The time needed should be obtained
from the piloting, so that the respondent can understand their com-
mitment. Ensure that the time allowed is reasonably accurate.

Commonly, anonymity will not be necessary, although confidentiality may be advisable, in order to obtain fuller and more readily
given responses. The assurances can be given verbally but should be
confirmed in writing in the formal letter of request for response in
which the purpose and legitimacy of the research should be
explained. It is useful if the letter contains an explanation of the
research, the envisaged outcomes, benefits and purpose of the work
as well as an explanation of its role in a degree course etc.

Despite assurances of confidentiality, such as, '...any data provided will be treated as confidential and used for the purposes of
this research only; the identity of respondents will not be revealed',
respondents may require further restrictions to apply concerning
publication of results. Such restrictions should be considered carefully as they could 'stifle' the work and its value. It is legitimate, of
course, to protect trade secrets, but 'vetting' of the contents of
research reports by 'commercial' organisations solely because they
have provided data should be avoided.

Vetting should be avoided as it is likely to restrict the report, by
suppressing or removing sections etc. and it may introduce bias.
Trade secrets etc. can be protected by confidentiality and/or anonymity measures, describing respondents as 'companies A, B and C', and/
or by restricting publication of the results etc. for a period of time or
by limiting their scope of circulation. If confidentiality is to be provided, ensure that it really is so – care in the report can be destroyed
by polite naming of respondents in 'Acknowledgements'.

As case study type data collections are in-depth, they are more
likely to encounter commercially sensitive issues. Hence, extra care
may be necessary to ensure confidentiality. As it is likely to be more

problematic to obtain the data, so cultivation of contacts can pay dividends. Sections of the research, and report, may have to be treated differently where the issues tackled vary in sensitivity, and hence in their confidentiality requirements.

Irrespective of the approach to data collection, piloting of the collection is vital. Executed well, amongst helpful, informed and appropriate respondents, it will reveal flaws in the data collection method and parameters (such as time required for the collection) can be determined. Modification and re-piloting will pay dividends by enhancing the rate of responses and quality of data obtained for analysis. Questions must be clear and precise and *not superfluous*.

Example

If you send a questionnaire by post to Ms A.Y. Lai, City Construction, Beijing, it is rather silly to have:

Question 1:   name of respondent
Question 2:   organisation's name and address

as they are obviously known and their inclusion will have the likely effect of upsetting the prospective respondent by indicating a lack of sense, sensitivity, attention to detail and courtesy by the researcher. 'How many other questions are unnecessary?' might be asked.

Questions must be clear and easy to answer.

Example
Instead of asking the following questions:

● Do you prefer tea or coffee?
● How many people did your organisation employ in 1994?

It would be preferable to ask:

(1)   Please tick which of the following beverages you prefer:

    (a)   tea   ☐
    (b)   coffee   ☐
    (c)   no preference ☐

(2)   Please tick the appropriate box to indicate how many people
      your organisation employed (on average) during calendar year
      1994.

      0–9   10–39   40–99   100–199   200–499   500–999   1000+
       ☐   ☐   ☐   ☐   ☐   ☐   ☐

In the original version of question (1), unless the respondent replied
'tea', 'coffee' or 'don't know', the answer would be ambiguous –
consider what 'yes' might indicate. The original question (2) suggests
that an answer to the nearest one employee is required, which is unlikely to be the case; the revised response indicator notes the level
of accuracy appropriate. In a large organisation, the exact number of
employees may vary from day to day, it may not be known exactly
by the respondent and it might be difficult to discover precisely.
Hence, a practical level of detail appropriate for the research and readily available to the respondents is required. All questions must be
material to the research, not just of passing interest. The '20-question'
guide is often quite useful and about the right number. Clearly, certain research projects must collect all possible data in order to search
for patterns etc., but in a constrained or quantitative research project,
such an approach is unlikely to be employed.

The more broad a concept is, the greater is the likelihood that it
encompasses underlying dimensions, each of which reflects a different, important aspect of the overall concept. Lazarsfeld (1958) suggested a four-stage process:

- Imagery,
- Concept specification (dimensions),
- Selection of indicators,
- Formation of scales/indices (for measurements).

Theory and literature may be helpful in identifying the dimensions;
a priori establishment of the dimensions is very useful in forming indicators of the concept. Factor analysis is useful in examining the relationships between the dimensions and how they relate to the concept.

Due to problems with collecting original data, researchers increasingly 'pool' data – data are collected by groups of researchers who

wish to investigate different but related topics – a common data set can represent an efficient compromise. Data collected by others can be used provided full details of the data collection methods, samples etc. are available as well as the data themselves. Published data are a good source and should be used as much as possible, not only for savings of time and cost, but also due to availability and convenience – such data can be used for a preliminary or background study – perhaps to provide the context for a detailed study involving collection of specific, original data.

If the particular data desired cannot be obtained, it is likely that surrogate measurements can be made. If treated with care, such measurements can yield very good approximations to what was sought originally, but do note the differences in the measurements and any changes in assumptions necessary.

So, data form the essentials of a research project; after collection and analysis, their interpretations must begin to yield meaning in the context of theory and literature. For postal questionnaires, a date for commencement of analysis should be set to allow adequate time to obtain responses, including issuing polite reminders. A date for receipt of responses by the researcher should be specified clearly on the questionnaire and in a covering letter, usually 2–3 weeks from respondents' receipt of the questionnaire; at that time, non-respondents can be given reminders by post, telephone, fax or email. However, if only non-respondents are targeted, this means that the researcher has a referencing mechanism and so responses are not anonymous. The date for commencement of analysis should be adhered to, and any responses received after that date excluded.

Moser and Kalton (1971) note six primary conditions for postal questionnaires, apart from non-responses:

- '. . . the questions . . . [must be] . . . sufficiently simple and straight-forward to be understood with the help of printed instructions and definitions . . . .
- . . . the answers . . . have to be accepted as final . . . . There is no opportunity to probe . . . .
- . . . are inappropriate where spontaneous answers are wanted . . . .
- . . . the respondent . . . can see all the questions before answering any of them, and the different answers cannot therefore be treated as independent.

- ...[the researcher]...cannot be sure that the right person completes the questionnaire.
- ...there is no opportunity to supplement the respondent's answers by observational data.'

They note that 'Ambiguity, vagueness, technical expressions and so forth must be avoided.... A mail questionnaire is most suited to surveys whose purpose is clear enough to be explained in a few paragraphs of print...and in which the questions require straightforward and brief answers'.

As in any survey, non-responses present a problem, not just because they reduce the size of the sample which can be analysed, but more notably because they may represent a body of opinion which, although unknown, may be significantly different from that which has been expressed by those who did respond. It may be the case also that the responses given by those for whom a follow-up reminder was necessary to obtain their responses form another group with a 'cluster' of opinions; however, analysis can reveal such clusters. Hence, the need for keeping good records of how, and when, responses were received.

However non-responses are dealt with – initial follow-up reminders, re-surveys etc. – note that, 'Non-response is a problem because of the likelihood – repeatedly confirmed in practice – that people who do not return questionnaires differ from those who do.... It has also been shown frequently that response level is correlated with interest in the subject of the survey.... If the response rate is not high enough to eliminate the possibility of serious bias, care must be taken to judge the extent of the unrepresentativeness and to take account of it in making the final estimates' (Moser and Kalton, 1971). Further, they note a tendency for there to be upward bias in terms of both educational and social level of those who do respond to general mail questionnaire surveys.

Clearly, there is no single best way of dealing with non-responses. Thus, response levels must be noted, along with the results obtained from the responses. They should be grouped as appropriate, as well as overall. Discussion of the impact of non-responses should be noted clearly as such. Separate what has been found out via the responses from what can be postulated reasonably to allow for non-responses.

For interview surveys, the issue of non-responses can be dealt with as the sample of respondents is being assembled. Interviews may be used to obtain greater 'depth' following a postal questionnaire survey or/and to obtain information about non-responses to the questionnaire. However, interviews may be subject to various sources and types of error and bias. Some have been noted earlier; those of the interviewer wishing to obtain support, those of interviewees wishing to 'please' the interviewer etc. It is important to remember that interviewing is a human, social process, however it is executed, and as such, it is subject to the interactions which occur between the participants. If the participants like each other, the process will be different and the responses may be different from a situation in which the participants dislike each other, even if it is only a first impression. The subject matter can influence such personal interaction too.

Thus, as well as training interviewers to ask questions and probe objectively, they must be able to record the situation of the interview accurately as well as the interviewees' responses. Cannell and Kahn, in Lindzey and Aronson (1968), note three conditions necessary for successful interviews:

- *accessibility*, to the interviewee of the information required,
- *cognition*; the interviewee's understanding of what is required,
- *motivation*, of the interviewee to answer the questions accurately.

Atkinson (1967) distinguishes three types of questions:

- factual,
- knowledge,
- opinion.

Opinion questions are the most sensitive category; factual are least sensitive. The more sensitive the category of questions, the more important it is that the questions are not perceived by the respondent to be 'threatening'. A 'threatening' question reduces the response rate to individual questions, if not to the entire interview or questionnaire.

Kahn and Cannell (1957) note five primary reasons for inadequate responses:

- partial responses,
- non-responses,
- irrelevant responses,

- inaccurate responses,
- verbalised response problem, where the respondent gives a 'reason' for not answering.

Many of the possible causes of inadequate or non-responses (including 'threatening' questions) should be removed by good piloting. Piloting should ensure also that the time required of interviewees is not unreasonable and that probing is being employed appropriately. Further, it will indicate whether the method of recording both responses and interview 'situations' etc. is adequate.

A particular feature of piloting is that it should be followed through with initial analysis, production of results etc. so that the data provision and use are checked thoroughly.

Bryman and Cramer (1994, p. 64) note that, 'if a question is misunderstood by a respondent when only one question is asked, that respondent will not be appropriately classified; if a few questions are asked, a misunderstood question can be offset by those which are properly understood'.

Two further issues which concern data are:

- reliability
- validity.

Reliability concerns the consistency of a measure. *External reliability* concerns the consistency of a measure over time – often measured by re-test reliability – ensuring that the tests are sufficiently far apart in time that the respondents do not answer the second test by recalling their responses to the first. Any events which have occurred between the occasions of the tests should be taken into account in examining the results of the test and retest.

Internal reliability concerns whether each scale is measuring a single variable; whether the items which constitute the scale are internally consistent. *Split half reliability* may be employed – the items in the scale are split into two halves and the respondents' scores in the two halves are examined for consistency.

Validity concerns how well a measure does measure the concept it is supposed to measure. *Concurrent validity* measures variables on which people differ and determines whether the measures of the

variables are consistent, including according with theory and literature. *Predictive validity* is where the variables are measured sequen-
tially, with a reasonable time separator; to determine whether the
supposed predictive measure is a good predictor. (See also: experimental control – Chapter 4; validities – Chapter 5.)

The concept of statistical sampling concerns obtaining a sample
which is adequate to represent the population under investigation
with sufficient confidence. Even where non-statistical sampling
methods are employed, such as a single case study of a construction
project using ethnomethodology to investigate relationships, there is
likely to be a desire to consider any validity of the results beyond the
sample obtained.

Strictly, of course, unless the research is designed to obtain data,
and hence, results, which are valid beyond the confines of the data
or sample, the findings are valid for the subject data only. The
issues concern statistical inference, deduction and induction. No
matter how well a research is designed, it remains subject to the
data actually obtained as responses, experimental measurements etc.
in determining its validity, both internal and external. Hence, following the discussions of sampling, responses and measurement, the
issue of adequacy of data response remains.

Clearly, data collection should be statistically designed to provide
sufficient expected responses, so that the desired validity will be satisfied through sample selection and sizing, allowing for anticipated
response rates. For example, if 'small number' statistics are to be
avoided, a minimum of 32 usable responses is necessary; given an
expected response rate of 30% for a postal questionnaire, a sample
size of over a hundred, minimum 107, is necessary.

Especially for surveys (and notoriously, the enormously popular
postal questionnaires) it is important to consider response rates. As
well as numbers of responses obtained, in order to evaluate the validity of the data, one must consider the likelihood of non-responses
from various groups being different, due to the issues of non-
responses likelihood of being different. Indeed, responses which are
obtained after 'follow-up' requests may constitute an intermediate
group between immediate responses and non-responses.

Fowler (1984) noted 'there is no agreed-upon standard for a minimum acceptable response rate'. There are a variety of perspectives to
evaluate in determining the responses to be sought; these relate to the

population, the sample, the tests and, most especially, the validity and applicability of the findings. Rigorous calculations and judgements are required.

Summary

This chapter has addressed a variety of issues concerning collection of data. In particular, the issues of sampling have been examined – size and structure. The various types of data have been discussed and the tests to which they may be subjected. Approaches to elicit data from respondents have been outlined along with the necessity to preserve confidentiality and anonymity in some instances. The issue of response rates is important and must be taken into account when deciding the size of sample. Piloting is vital to ensure data provision by respondents is easy and the requirements are clear.

References

Ahmad, I. & Minkarah, I. (1988) Questionnaire survey on bidding in Construction, *ASCE Journal of Management in Engineering Divisions*, 4(3), July, 229–243.

Atkinson, J. (1967) *A Handbook for Interviewers: a Manual for Government Social Survey Interviewing Staff, Describing Practice and Procedures on Structured Interviewing*. Government Social Survey No. M136, HMSO, London.

Bell, J. (1993) *Doing Your Research Project*, Open University Press, Milton Keynes.

Bryman, A. & Cramer, D. (1994) *Quantitative Data Analysis for Social Scientists*, revised edn. pp. 62–70, Routledge, London.

Bryman, A. & Cramer, D. (1994) *Quantitative Data Analysis for Social Scientists* (revised edn), Routledge, London.

Cleland, D.I. & King, W.R. (1983) *Systems Analysis and Project Management*, 3rd edn, McGraw-Hill, Singapore.

Fiedler, F.E. (1967) *A Theory of Leadership Effectiveness*, McGraw-Hill, New York.

Fowler, F.J. (1984) *Survey Research Models*, Sage, London.

Kahn, R.L. & Cannell, C.F. (1957) *The Dynamics of Interviewing; Theory, Technique and Cases*, Wiley, New York.

Lazarsfeld, P.F. (1958) Evidence and inference in social research, *Daedalus*, **87**, 99–130.

Lindzey, E. & Aronson, G. (1968) *The Handbook of Social Psychology, Vol. 2, Research Methods*, 2nd edn, Addison-Wesley, Reading, Massachusetts.

Moser, C.A. & Kalton, G. (1971) *Survey Methods in Social Investigation*, 2nd edn, Dartmouth.

Newcombe, R., Langford, D.A. & Fellows, R.F. (1990), *Construction Management 1: Organisation Systems*, Mitchell, London.

Osgood, C.E., Suci, G.J. & Tannenbaum, P.H. (1957) *The Measurement of Meaning*, University of Illinois Press, Illinois.

Rowlinson, S.M., Ho, T.K.K. & Po-Hung, Y. (1993) Leadership style of construction managers in Hong Kong, *Construction Management and Economics*, **11**(6), 455–465.

Shash, A.A. (1983) Factors considered in tendering decisions by top UK contractors, *Construction Management and Economics*, **11**(2), 111–118.

Yeomans, K.A. (1968) *Statistics for the Social Scientist 2: Applied Statistics*, Penguin.

Chapter 7

Data Analysis

The objectives of this chapter are to:

- introduce logical procedures for **analysing data**;
- demonstrate the value of **plotting data** in various ways;
- discuss some primary **statistical methods** for analysing data – **non-parametric tests** and **parametric tests**.

Analysing data

It has been noted that the choice of data collected should be determined by the outputs required from the research, given constraints of practicality. One consideration, given the systems perspective of Chapter 6, is the analysis of the data which is to be undertaken. Unfortunately, it has become all too common, especially for new and enthusiastic researchers, to plunge into the most complex statistical techniques they can find, often using computer packages in a 'black box' fashion, only to 'emerge' some while later rather bemused.

The preferable approach is to consider, evaluate and plan the analysis in a similar way to planning the whole research project. Geddes (1968), the 'father of town planning', advocated the method of:

- survey
- analyse
- plan.

A sensible way to ensure that the methods selected are appropriate.

Not all research projects yield data which are suitable for statistical analyses, and even those which do may require only simple manipulations of small sets of data. Computing helps but is not essential – it makes calculations quicker and easier – an advantage which can cause major problems too. Problems can occur as packages are, increasingly, 'user friendly', some to the extent that the researcher may not be sufficiently aware of the statistical bases and assumptions of the tests. In the early days of the particular package called Statistical Package for the Social Sciences (SPSS), an expert user obtained only a little printout, as the appropriate tests had been performed on the data; a novice user received a 'mountain' of printout as everything available had been performed on the data – the user was left with the problem of sorting through the printout to extract useful elements. So, 'user friendly' programs are very helpful, but can cause the 'black box' syndrome; it is important to remember that, to be useful, tests must be valid and understood.

No matter what is the nature of the data collected, it is appropriate to begin analysis by examining the raw data to search for patterns. Of course, a pattern or a relationship may be expected from the review of theory and literature – one may have been hypothesised. Alternatively, for fundamental studies where theory and literature do not exist to any great degree, the search for patterns and relationships in the data and the identification of major variables may constitute the total analysis for the research project.

For data sets in topics which have an extensive body of theory and literature, it is good practice to search the data, *with an open mind*, for themes and categories. Of course, usually, such scrutiny will serve to confirm the themes and categories etc. found in the theory and literature, but the researcher must be prepared to discover differences in the data from what theory and previous findings suggest will occur. Societies are dynamic, so changes over time should be expected, and methods of analysis must be sufficiently rigorous to detect them.

Qualitative data can be difficult and laborious to analyse – they must be handled systematically; a requirement which is easier with quantitative data. Categorisation of qualitative data may rely on the researcher's opinion; it may be useful to construct a set of guidelines initially, and to confirm or amend and supplement them on a 'first pass' of the data. A 'second pass', using the completed categorisation, will ensure that all of the data, especially the data considered early in the first pass, are categorised consistently. For large sets of data, a 'piloting' exercise using a sample may serve as the first pass. Essentially, the approach arises as part of grounded theory. In such an exercise, it is necessary to consider each transcript so that the contexts of words are not lost. For example, the word 'tap' has a variety of meanings in England; equally, slang terms vary in meaning.

Thus, for much 'human-oriented' research in construction, the objective is to analyse peoples' behaviours, including their causes and consequences, as manifested in actions and symbols – notably oral and written language. Some data may be documented outside of the research, such as 'newspaper' articles, others may be obtained specifically (e.g. questionnaires), and others may be collected through researcher observation, during interviews, notes of meetings, notes of 'shadowings' etc. Respondents' diaries may be obtained for analysis, although to avoid bias, it is preferable to obtain diaries which have been kept by respondents as a matter of course or daily routine in their daily work. Where data are in an oral form, such as tape recordings of interviews, usually, it is best to transcribe the data prior to analysis, in order to aid clarity of data and consistency of analysis.

In most contexts, visual aids and diagrams can be extremely helpful in analysing data, as patterns and relationships often emerge. Such diagrams should comprise (as near as is practical) the raw data; this is relatively simple for quantitative data but will be the result of the initial scrutinies where categories of qualitative data are required. For 'second hand' data, such as statistics published by government, the raw data may not be available, or they are likely to be inconvenient to access, so the published data must be used.

Although the patterning revealed by examination of quantitative data may be quite straightforward, two considerations involving potential difficulties may occur. The first is that theory and literature, by advancing relationships between variables via hypotheses, may lead to other possible relationships not being considered. The second is to ensure that, especially if using a computer, the data have been input correctly. Usually data have to be coded. Whilst this tends to be simple and obvious for quantitative data, coding may distort, or be part of the analysis, of qualitative data. Ensure the coding is both easy to use and understand and is of an appropriate level of detail. Too little detail will yield 'conglomerate categories' which do not reveal meaning. Too much detail will not only produce allocation problems, but also yield so many categories that there may be almost a category for each item of data, rendering analysis unwieldy.

Moser and Kalton (1971) note that the sets of data from each respondent should be subjected to an editing process before coding. The editing should check the data sets for completeness – so that any gaps may be filled, if possible; for accuracy – to check or verify any apparent inconsistencies; for uniformity – so that responses are in the same form for coding, notably where interviews have been employed or/and semantic responses may be used to produce a frame or keyword set of response contents: an elementary form of content analysis.

As noted above, many analyses of qualitative data concern searching the data for patterns of various types, so that hypothesised relationships can be established for subsequent investigation and testing – perhaps, by more quantitative methods. In seeking patterns, two main approaches may be employed, either individually or together, to search the data for patterns (e.g. as in grounded theory), or to employ theory and literature to suggest likely 'rational' patterns; however, the latter approach, if adopted alone, may result in the researcher's missing new, and potentially important, relationships in the data – an 'open' mind, as free as possible from preconceptions, is likely to be most appropriate and revealing.

Many qualitative approaches are not subject to particular analytic techniques with prescribed tests, as is common in quantitative analyses, such as conversation and discourse analyses. Instead, they involve scrutiny of transcribed texts of discussions, statements etc. so that not only is the content analysed but the linguistic context is considered, in order to establish the meanings, intentions, interpretations etc. of the people concerned. Hence, the researcher must develop sensitivity to the people, their language and the way in which language may be used. Discourse analyses involve many readings of the texts being analysed, so that iterative formulation, testing and revision of hypotheses concerning the discourses in the texts may occur. For this reason the context of the texts of the discourse is important as an indicator of possible or likely purpose.

Content analysis may be employed, at its most simplistic, to determine the main facets of a set of data, by simply counting the number of times an activity occurs, a topic is mentioned etc. However, even for such apparently straight-forward analysis, awareness and interpretations by the researcher are likely to be necessary – a number of actions, although different, may be very similar in purpose; several words or phrases etc. may have very similar meanings – and so the boundaries of categories must be established to a sufficient, but not overwhelming, extent for the analysis. For many content analyses, it is important to have a sound theoretical basis to assist development and testing of hypotheses – such as non-verbal behaviours of people in meetings. This is so that actions, such as those which indicate aggression, nervousness etc., in that society can be identified. Clearly, virtually identical behaviours can have different meanings depending on the contexts so, here too, the situation should be considered holistically for analysis. Thus, once the categories of data have been established, a content analysis will yield quantitative data for each content category. Some analyses may be required to yield such descriptive results only; others may wish to continue to investigate relationships – using correlations or more multidimensional analyses.

The initial step in content analysis is to identify the material to be analysed. The next step is to determine the form of content analysis to be employed – qualitative, quantitative or structural; the choice is dependent on, if not determined by, the nature of the research project. The choice of categories also will depend upon the issues to be addressed in the research if they are known.

In qualitative content analysis, emphasis is on determining the meaning of the data. Data are given coded allocations to categories and groups of 'respondents' from whom the data were obtained are fitted to these categories, so that a matrix of categorised data against groups are obtained. Statements etc. can be selected from each cell of the matrix to illustrate the contents of each of the cells. As in any allocation mechanism, the categories should be *exclusive*, i.e. data assigned to one category only, and *exhaustive*, i.e. categories cover the research topic comprehensively.

Quantitative content analysis extends the approach of the qualitative form to yield numerical values of the categorised data – ratings, frequencies, rankings etc., which may be subjected to statistical analyses. Comparisons may be made and hierarchies of categories may be examined.

Structural content analysis concerns determination and examination of relationships between categories of data and between groups where this is appropriate. The rules used to determine relationships will depend on the aims of the research project.

Irrespective of the form of content analysis employed, there may be a tendency only to consider what is mentioned in the transcript or data; in some cases, what is omitted is of great importance and astute researchers should consider such omissions. Further, not just categories of data but combinations of categories may be important too.

The structures of qualitative data may be investigated using Multi-dimensional Scalogram Analysis – MSA (Lingoes, 1968). MSA facilitates setting research subjects (people in a survey or respondents) and the variables under consideration, to be shown on one diagram. Given the research subjects and the dimensions of the variables, a matrix can be constructed. By grouping subjects' scores against the variables, the numbers in the cells of the matrix can be ascertained. Scores for the variables are best kept simple (say in the range 1–4). Diagrams are produced which represent the results to aid comparisons between subjects. This is discussed and demonstrated by Wilson (1995).

To a large extent, the nature of diary data depends on what, if any, structuring of the diary is imposed; especially for 'freely composed' diaries, various analytic techniques are available but, initially, a content analysis is likely to be helpful. Once the contents of the diaries have been categorised, and hence, given some common structure, it may be appropriate to proceed with more quantitative analyses, as well as qualitative and descriptive ones.

The purpose of analysing the data is to provide information about variables and, usually, relationships between them. Hence, as research in a topic becomes more extensive, quantitative studies may be undertaken to yield statistical evidence of relationships and their strengths; statistics are useful in determining directions of relationships (*causalities*) when combined with theory and literature.

However, the purpose of analysis is to provide evidence of relationships and to aid understanding; in a context of management, it is to support decision making – hence, the importance of *inference*. Inference is what follows logically from the evidence, and it is important to know how valid those inferences are. Popper (1989, pp. 317–318) notes that, 'a rule of inference is valid if, and only if, it can never lead from true premisses to a false conclusion'. A summary of some quite simple statistical techniques which are used extensively in analyses of research data follows. Most computer statistics packages can perform the numerical manipulations but *the researcher must understand what is being done*. Beware the idiot machine!

Plotting data

Once the data have been collected, it is helpful to produce a diagram or graph of those data. Such plots will help to indicate the natures of distributions of the data and relationships between them such that appropriate statistical techniques, if any, may be employed in analysis. Not all data lend themselves to plotting in the form of a graph. For dichotomous variables, such as 'yes' or 'no' responses, cross-tabulations, or contingency tables, are used to detect patterns. The next step is to undertake a statistical analysis, the most usual of which is the χ^2 (chi-square) test.

A table of desired and actual responses, such as to questionnaires sent and received back, noting proportion usable for the research, is useful to demonstrate the sampling attempted and realised. Use both actual numbers and percentages to convey maximum information. Having depicted the data (responses) being considered, focus can fall on analyses. Analysis examines responses so that patterns and relationships between variables can be discovered and quantified, *with theory helping to explain causation*. Consider, for example, percentage changes in costs observed in a sample of projects. The data are summarised in Fig. 7.1.

% change	−10 to −5.1	−5 to −0.1	0 to 4.9	5 to 9.9	10 to 14.9	15 to 19.9	20 to 24.9	25+
No. projects	0	3	10	21	14	6	1	0

Fig. 7.1 Frequency distribution.

Now consider the diagrammatic representation shown in Fig. 7.2 and Fig. 7.3 of the data in Fig. 7.1.

Histograms have the particular property that the area of each rectangle represents the proportion of the number of observations in that class; this property does *not* apply to a *bar diagram*.

It is quite simple to convert a frequency polygon into a *frequency curve*. A frequency curve gives an indication of the shape of the sample distribution, and consequently of the population distribution, provided good sampling techniques have been used.

An *ogive* (Fig. 7.4) is a form of cumulative frequency distribution curve. Using the upper boundary of the class intervals yields the 'more than' cumulative curve.

Presentation of 'raw' data provides the greatest detail. However, even if ordered in some way, the data may not be easy to interpret. In presenting data, it is common for detail to be sacrificed, so that intelligibility is improved by the use of tables and diagrams. In any event, it is helpful if the 'raw' data are presented in an appendix to the research report; data are valuable for further studies, tests etc. as well as for checking and verification of the instant analysis and results.

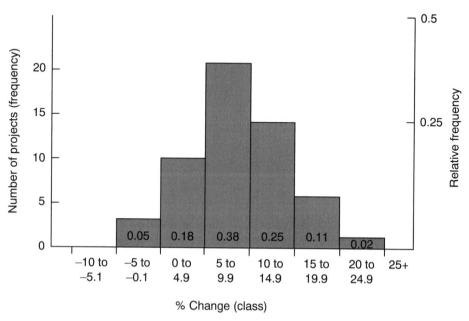

Fig. 7.2 Bar diagram (relative frequencies may not sum to 1.0 due to rounding).

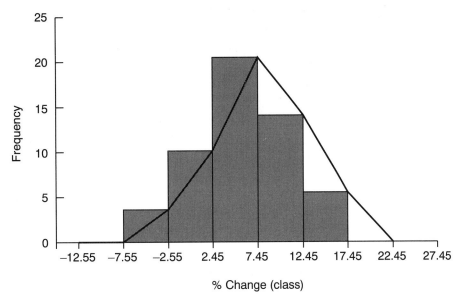

Fig. 7.3 Frequency polygon or *histogram*.

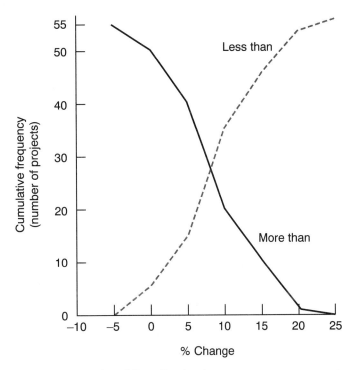

Fig. 7.4 'More than' and 'less than' ogives.

Statistical methods

Some common statistical methods used in data analysis are discussed in this chapter:

- non-parametric tests
  - sign test
  - rank sum test
  - chi-square test
  - goodness of fit
- parametric tests
  - *t* test
  - ANOVA (analysis of variance)
- regression and correlation
- times series
- index numbers

Non-parametric tests

Non-parametric tests are distribution-free, and so are more flexible in application.

Sign test

The sign test examines paired data using positive $\oplus$ and negative $\ominus$ signs.

Example: The Sign test
A sample of architects are asked to rate the performance of two types of roofing tiles, A and B; the scorings are: excellent = 5, to very poor = 1, on a 5-point semantic scale.

Customer No.	1	2	3	4	5	6	7	8	9	10	11	12
(1) Type A	5	4	5	2	1	2	3	2	5	5	3	1
(2) Type B	3	4	2	1	3	3	4	5	3	4	1	5
Sign [(1) − (2)]	+	0	+	+	−	−	−	−	+	+	+	−

No. +ve (p)	6
No. −ve (q)	5
No. 0	1
Total	12

H_O: $p = 0.5$ (there is no difference between tiles A and B)
H_A: $p \neq 0.5$ (there is a difference between tiles A and B)

excluding zeros:

$n = 11$
$\bar{p} = 6/11$ (proportion of 'successes')
$\bar{q} = 5/11$ (proportion of 'failures')

for 'no difference', $p\,H_O = q\,H_O = 0.5$

Standard error of the proportion:

$$\sigma_{\bar{p}} = \sqrt{\frac{pq}{n}} = \sqrt{\frac{(0.5)(0.5)}{11}}$$

$$= \underline{0.151}$$

As H_A: $p \neq 0.5$ (i.e. concerned with larger OR smaller), a 2-tailed test is required.

At 0.05 level of significance and as np and $nq \nless 5$, the normal distribution approximates to the binomial, the z value for 0.475 (i.e. 0.5 minus $\frac{1}{2} \times 0.5$) of the area under one tail of the normal curve is 1.96, then:

$$p\,H_O + 1.96\sigma_{\bar{p}} = 0.5 + (1.96)(0.151)$$

and

$$p\,H_O - 1.96\sigma_{\bar{p}} = 0.5 - (1.96)(0.151)$$

So, the range of acceptance is:

$$0.204 \rightarrow 0.796$$

The sample proportion,

$$\bar{p} \left(= \frac{6}{11} \right) = 0.545$$

As $0.204 < 0.545 < 0.769$, there is *no difference* in the architects' perceptions of the tiles (H_O is accepted).

Rank sum tests

Rank sum tests are used to test whether independent samples have been drawn from the same population.

The *Mann–Whitney* U *test* is used when there are two samples, and the *Kruskal–Wallis* K *test* is used when there are three samples or more.

Example: Mann–Whitney U

Consider rents of a particular type of building in two locations X and Y; the locations are quite close to each other; rents are expressed in $ per m^2 of floor area.

Rents in Location X ($/m^2)	Rents in Location Y ($/m^2)
30	39
32	42
40	54
50	52
55	36
47	46
48	45
38	37
41	33
53	51

Rank	Rent	Location	Rank	Rent	Location
1	30	X	11	45	Y
2	32	X	12	46	Y
3	33	Y	13	47	X
4	36	Y	14	48	X
5	37	Y	15	50	X
6	38	X	16	51	Y
7	39	Y	17	52	Y
8	40	X	18	53	X
9	41	X	19	54	Y
10	42	Y	20	55	X

n_1 = number of buildings in sample 1
n_2 = number of buildings in sample 2
R_1 = sum of ranks in sample 1 (in X locations)
R_2 = sum of ranks in sample 2 (in Y locations)

From the example:

$n_1 = 10$, $n_2 = 10$, $R_1 = 106$, $R_2 = 104$

U = Mann–Whitney U statistic

Using the standard form of the Mann–Whitney U test:
 The U-statistic is a measure of the difference between the ranked observations of the two samples:

$$U = n_1 n_2 + \frac{n_1(n_1 + 1)}{2} - R_1$$

$$= 10 \times 10 + \frac{10(11)}{2} - 106$$

$$= \underline{\underline{49}}$$

H_O: samples are from the same population
H_A: samples are from different populations

If H_O applies, samples are from the same population and the U-statistic has a sampling distribution described by:

$$\mu_u = \frac{n_1 n_2}{2}$$

$$= \underline{\underline{50}}$$

μ_u = mean
z-value = confidence level required
 α = level of confidence

$$\text{standard error, } \sigma_u = \sqrt{\frac{n_1 n_2 (n_1 + n_2 + 1)}{12}}$$

$$= \sqrt{\frac{(10 \times 10)(21)}{12}}$$

$$= \underline{\underline{13.23}}$$

So H_O: $\mu_1 = \mu_2$
 H_A: $\mu_1 \neq \mu_2$

$\alpha = 0.05$ (i.e. 95% confidence level)
z-value (using normal distribution) of $0.475 = \underline{1.96}$

$$\text{Limits: } \mu_u + 1.96\sigma_u = 50 + (1.96)(13.23)$$
$$= \underline{75.93}$$
$$\mu_u - 1.96\sigma_u = \underline{24.07}$$

As $24.07 < 49.0 < 75.93$, H_O is accepted. [i.e. $(\mu_u - 1.96\sigma_u) < u < (\mu_u + 1.96\sigma_u)$]

(*Note*: If items of data have equal values, the rank assigned to each one is averaged.)

Example: Kruskal–Wallis K
Tests have been carried out on three types of dumper trucks to determine the distance each travels on site using one gallon of fuel. The results, in miles, are:

Truck type A   6.0 6.8 5.7 5.2 6.5 6.1
Truck type B   5.6 5.9 5.4 5.8 6.2 7.0 5.1
Truck type C   5.0 6.3 5.3 6.4 6.6 5.5 6.7

Rank	Distance	Type	Rank	Distance	Type
1	5.0	C	11	6.0	A
2	5.1	B	12	6.1	A
3	5.2	A	13	6.2	B
4	5.3	C	14	6.3	C
5	5.4	B	15	6.4	C
6	5.5	C	16	6.5	A
7	5.6	B	17	6.6	C
8	5.7	A	18	6.7	C
9	5.8	B	19	6.8	A
10	5.9	B	20	7.0	B

Using the standard form of the Kruskal–Wallis K statistic:

$$K = \frac{12}{n(n+1)} \sum \frac{R_j^2}{n_j} - 3(n+1)$$

where:

$K =$ Kruskal–Wallis K statistic

$n_j =$ no. of items in sample j

$R_j =$ sum of the ranks of the items in sample j

$k =$ number of samples

$n = n_1 + n_2 + \cdots + n_k$

$=$ the total number of observations in all the samples

Type A	Rank	Type B	Rank	Type C	Rank
5.2	3	5.1	2	5.0	1
5.7	8	5.4	5	5.3	4
6.0	11	5.6	7	5.5	6
6.1	12	5.8	9	6.3	14
6.5	16	5.9	10	6.4	15
6.8	19	6.2	13	6.6	17
		7.0	20	6.7	18
	$\overline{69}$		$\overline{66}$		$\overline{75}$

$$K = \frac{12}{20(20+1)} \left[\frac{(69)^2}{6} + \frac{(66)^2}{7} + \frac{(75)^2}{7} \right] - 3(20+1)$$

$$= 0.02857[793.5 + 622.3 + 803.6] - 63$$

$$= \underline{0.408}$$

According to Levin and Rubin (1990, p. 609), the K-statistic can be approximated by a chi-square distribution when all the sample sizes are at least 5. The number of degrees of freedom is $k - 1$.

H_O: $\mu_1 = \mu_2 = \mu_3$

H_A: μ_1, μ_2, μ_3 are *not* equal

$\alpha = 0.05$

From tables of the chi-square distribution; with 2 degrees of freedom and 0.05 of the area in the right hand tail, $\chi^2 = 5.991$.

As the calculated value of K is less than the tabulated value of χ^2, the sample lies within the acceptance region and so, H_O should be accepted; there is no difference between the trucks.

Chi-square (χ^2) test

The *chi-square test* is used to compare observed and expected frequencies of a variable which has three or more categories, to test whether more than two population proportions can be considered to be equal. Generally, the χ^2 distribution should *not* be used if any cell contains an expected frequency of less than 5.

Example

The numbers of male and female workers are noted over three construction sites. The researcher wishes to investigate possible sex discrimination between the construction sites, and so wishes to know if the data provide any evidence.

H_0: there is no difference in the proportion of female workers employed on each site.

	Site A	Site B	Site C	Total
Male	52	48	60	160
Female	13	15	12	40
Total	65	63	72	200

Normalising the data (rounding to whole numbers):

	Site A	Site B	Site C	Total
% Male	33	30	37	100
% Female	33	38	29	100
% Total	33	31	36	100

Hence, the question is, given a sample of 200 workers on the construction sites, is it reasonable for the female workers to be distributed $A = 13$; $B = 15$; $C = 12$, if there is no sexual discrimination between those sites?

Expected female workers (rounded):

$$\text{Site A: } \frac{65}{200} \times 40 = 13$$

$$\text{Site B: } \frac{63}{200} \times 40 = 13$$

$$\text{Site C: } \frac{72}{200} \times 40 = 14$$

Expected male workers:

$$\text{Site A: } \frac{65}{200} \times 160 = 52$$

$$\text{Site B: } \frac{63}{200} \times 160 = 50$$

$$\text{Site C: } \frac{72}{200} \times 160 = 58$$

$$\chi^2 = \sum \frac{((f_o - f_e)^2)}{f_e}$$

where:

f_o = observed frequency
f_e = expected frequency

so

$$\chi^2 = \left[\frac{(52-52)^2}{52} + \frac{(48-50)^2}{50} + \frac{(60-58)^2}{60} \right.$$
$$\left. + \frac{(13-13)^2}{13} + \frac{(15-13)^2}{13} + \frac{(12-14)^2}{14} \right]$$

$$= [0 + 0.08 + 0.067 + 0 + 0.308 + 0.286]$$

$$= \underline{0.741}$$

The number of degrees of freedom of a χ^2 distribution is:

$$(\text{no. of rows} - 1)(\text{no. of columns} - 1)$$

Hence, in the example, the degrees of freedom are:

$$(2 - 1)(3 - 1) = 2$$

The tabulated value of χ^2 with 2 degrees of freedom and $\alpha = 0.5$ is 5.991. α = level of confidence.

Thus, as the calculated χ^2 is less than the tabulated value, the null hypothesis cannot be rejected; there appears to be no sexual discrimination in worker employment between the construction sites.

Goodness of fit

The goodness of fit of the data to a theoretical distribution is examined by the *Kolmogorov–Smirnov* test. The χ^2 test can be used for this purpose.

The Kolmogorov–Smirnov statistic, D_n, is the maximum value of the absolute deviation of $f_e - f_o$, where f_e and f_o are expected and observed relative cumulative frequencies. Critical values are tabulated such that if the calculated value of D_n is less than the tabulated value, the null hypothesis that the sample accords with the distribution postulated cannot be rejected and is thus accepted.

Parametric tests

Parametric tests assume that the distribution is known, or that the sample is large, so that a normal distribution (see Fig. 5.3) may be assumed; equal interval or ratio scales should be used for measurements.

t test

The *t* test is used to determine if the mean of a sample is similar to the mean of the population.

$$t = \frac{\bar{x} - \mu}{\hat{\sigma}_{\bar{x}}}$$

where:

$\hat{\sigma}_{\bar{x}}$ is the estimated standard error of the mean.

Degrees of freedom applicable are $(n - 1)$.

t_{tab} = value taken from tabulated t distribution curve

If $t_{calc} < t_{tab}$, the mean of the sample is not significantly different from the mean of the population.

The test may be used to examine the means of two samples:

$$t = \frac{\bar{x}_1 \sim \bar{x}_2}{\text{standard error of the difference in means}}$$

$$\bar{x}_1 = \text{mean of sample 1}$$
$$\bar{x}_2 = \text{mean of sample 2}$$
$$\bar{x}_1 - \bar{x}_2 = \text{difference between the means}$$

Analysis of variance (ANOVA)

Used to test the significance of differences among more than two sample means.

$$H_O: \mu_1 = \mu_2 = \cdots = \mu_n$$
$$H_A: \mu_1 \neq \mu_2 \neq \cdots \neq \mu_n$$

The method assumes that each sample is drawn from a normal population; each population has the same variance.

$$F = \frac{\text{between groups estimated variance}}{\text{within groups estimated variance}}$$

sample variance:

$$S^2 = \frac{\sum (x - \bar{x})^2}{n - 1}$$

Variance among sample means:

$$s_{\bar{x}}^2 = \frac{\sum (\bar{x} - \bar{\bar{x}})^2}{k - 1}$$

where:

$\bar{\bar{x}}$ is the ground mean (i.e. the arithmetic mean of all the values of all the samples)

k is the number of samples

As the standard error of the mean, $\sigma_{\bar{x}}$, is $\sigma/\sqrt{n}$, Levin and Rubin (1990, p. 439) show that the first estimate of the population variance, the between-groups variance, is:

$$\hat{\sigma}^2 = \frac{\sum n_j (\bar{x}_j - \bar{\bar{x}})^2}{k - 1}$$

$n_j = \text{number of items in sample } j$

The within group variance:

$$\text{sample variance,} \quad S^2 = \frac{\sum (x - \bar{x})^2}{n - 1}$$

The second estimate of the population variance, the within group variance, is:

$$\hat{\sigma}^2 = \sum \left(\frac{n_j - 1}{n_T - k} \right) S_j^2$$

where:

$$n_T = \sum n_j$$

As $F \rightarrow 1$, the likelihood that H_O is valid increases; as the value of F increases, the likelihood of H_O being valid decreases.

Degrees of freedom in the numerator: $(k - 1)$
Degrees of freedom in the denominator: $(n_T - k)$

Using tables of the F distributions and the appropriate degrees of freedom; if $F_{calc} < F_{tab}$, the null hypothesis should *not* be rejected.

To be valid, the F test can be applied to large samples only, $n \geq 100$ (Yeomans, 1968, p. 101).

Regression and correlation

Usually, regression and correlation are considered together in expressing a relationship between two variables: one or more known values, realisations of the independent variable; and the other unknown, the dependent variable. To keep research clear, it is advisable, at least in the beginning, to consider variables in pairs – one independent and one dependent.

Regression and correlation statistics establish only any *relationship* between the realised values of the variables which occur; they *do not establish causality*, that is the province of theory, evidence and logical reasoning, in the light of the statistics. Conventionally, the independent variable is plotted on the *x*-axis and the hypothesised dependent variable on the *y*-axis. Simple or linear regression considers straight line hypothesised relationships only.

The standard form equation for a straight line is:

$$y = a + bx$$

where a is the intercept of the line on the y-axis and b is the slope of the line.

So, given at least two data points on a scatter plot (a graph of the associated values of x and y), a regression line can be drawn.

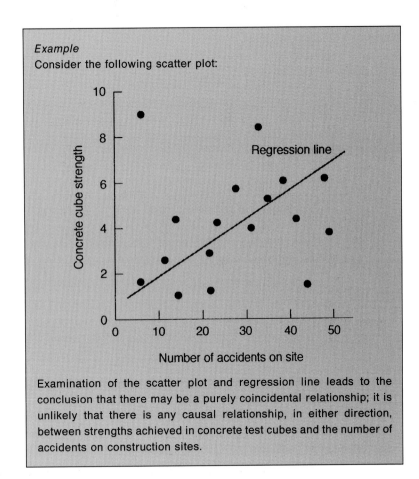

Example
Consider the following scatter plot:

Examination of the scatter plot and regression line leads to the conclusion that there may be a purely coincidental relationship; it is unlikely that there is any causal relationship, in either direction, between strengths achieved in concrete test cubes and the number of accidents on construction sites.

Note, however, that there is a positive relationship when the regression line has a positive slope, upwards from left to right.

The regression line, the line of 'best fit' through the data points, uses the criterion of least squares. Squaring the vertical distance of each data point from the regression line both magnifies errors and removes the possible cancelling effects of positive and negative distances. A regression line is used for estimation – there will be errors between the line and the actual, realised data points. As the line is used to estimate points on the y-axis, it is usual for the equation for a straight line of estimation to be:

$$\hat{Y} = a + bX$$

where: $\hat{Y}$ (Y-hat) are values on the y-axis estimated by the equation.

Least squares error, to determine the line of best fit, minimises $\sum (Y - \hat{Y})^2$.

Given a set of data points which relate the independent variable X and the (hypothesised) dependent variable Y, Levin and Rubin (1990; p. 491) note the equations to find the line of best fit to be:

$$b = \frac{\sum XY - n\bar{X}\bar{Y}}{\sum X^2 - n\bar{X}^2}$$

$$a = \bar{X} - b\bar{Y}$$

where:

b = slope of the line of best fit (estimate/regression) line
X = values of the independent variable
Y = values of the (hypothesised) dependent variable
$\bar{X}$ = mean of the values of X
$\bar{Y}$ = mean of the values of Y
n = number of data points (pairs of values of the variables X, Y)
a = Y-intercept.

The standard error of estimate measures the variability of the actual (realised) values from the regression line.

$$S_e = \sqrt{\frac{\sum (y - \hat{y})^2}{n - 2}}$$

Hence, the standard error of estimate measures the reliability of the estimating equation; analogous to standard deviation, it is a measure of dispersion. Levin and Rubin (1990; pp. 498–500) note:

$$S_e = \sqrt{\frac{\sum y^2 - a \sum y - b \sum xy}{n - 2}}$$

Assuming a normal distribution applies, the standard error of estimate exhibits the same properties as standard deviation, and can be used in the same way, for determining variability of predictions and confidence in them.

Regression assumes that the scatter of data points around the line of best fit is 'random', otherwise called *homoscedastic*. If there is a pattern to the scatter of the data points about the line, which shows the scatter to be different at different points, *heteroscedasticity* is present, and so regression is questionable. Strictly, homoscedasticity is where the error has constant variance; heteroscedasticity is where the variance of the error changes along the length of the regression line (Pindyck and Rubinfeld, 1981, p. 49).

Not all scatter plots suggest that straight lines are the best fit, so curve fitting may be appropriate. Fortunately, as for a straight line, there are standard forms of equation for various types of curves – usually determined by the approximate shapes of the curves, such as slope changes and any turning points. Use of computer packages is helpful for both linear and non-linear regression. The shape of the line and its nature should be detectable, along with any close alternatives, from observation of the scatter diagram, and in particular from the nature of the relationship suggested by any underpinning theory. Theory is important!

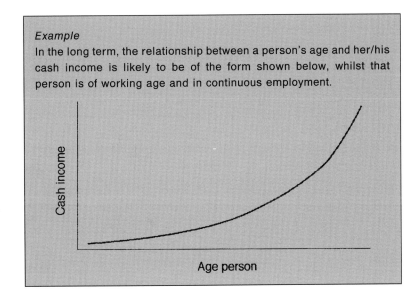

Example
In the long term, the relationship between a person's age and her/his cash income is likely to be of the form shown below, whilst that person is of working age and in continuous employment.

The line is an increasing exponential curve as income increases by an approximately constant percentage each year, so that the cash sum received per annum increases progressively. Economic theory has been used to determine the nature of the relationship. Collection of data and application of statistics would allow analytic checking and quantification of the relationship. Fortunately, most statistics packages, via simple commands, manipulations of data and setting of criteria, allow the program to determine the line of best fit.

The *coefficient of correlation*, r, identifies the degree and nature of the relationship between the two variables, from a perfect positive relationship (+1) to a perfect negative relationship (−1).

$$\text{i.e. } -1 \leq r \leq +1$$

$r = +1$ means that an increase in variable x is matched by an equi-proportional increase in y. If $r = 0$ there is no relationship; changes in the variables are quite independent of each other; they are random. However, it is common to wish to know how much of the change in the values of a dependent variable are caused, given the logic of the relationship, by a change in the values of the independent variable. The statistic required is the *coefficient of determination*, r^2.

If $r^2 = 0.81$ (i.e. $r = 0.9$), 81% of the changes in y are caused (explained) by the changes in x.

The coefficient of determination can be calculated using the following method:

A sample of data has variation about its own mean which, in terms of least squares error, is:

$$\sum (y - \bar{y})^2$$

Similarly, the variation of the data about the regression line is:

$$\sum (y - \hat{y})^2$$

The sample coefficient of determination is:

$$r^2 = 1 - \frac{\sum (y - \hat{y})^2}{\sum (y - \bar{y})^2}$$

The coefficient of determination measures the strength of a *linear* relationship between two variables.

Levin and Rubin (1990; p. 510) note a 'short cut method' for calculation of r^2 and most statistics packages can calculate r^2 directly:

$$r^2 = \frac{a \sum Y + b \sum XY - n\bar{Y}^2}{\sum Y^2 - n\bar{Y}^2}$$

If the true regression line for the population is given by $Y = A + BX$ and the line estimated from the sample is $\hat{Y} = a + bX$, the standard error of the regression coefficient can be used to test null hypotheses – for example, that the slope of the regression line is unchanged from what has been found in the past (the 'proportionate' relationship between Y and X). This is analogous to the use of standard deviation.

As B denotes the slope of the population's regression line, S_b denotes the standard error of the regression coefficient of b.

$$S_b = \frac{S_e}{\sqrt{\sum X^2 - n\bar{X}^2}}$$

$\Bigg(S_e$ is the standard error of estimate;

$$= \sqrt{\frac{\sum Y^2 - a \sum Y - b \sum XY}{n - 2}} \Bigg)$$

Using the t-distribution with $n - 2$ degrees of freedom, the limits for the acceptance region in this instance, showing that B is unchanged, are:

Upper: $B + t(s_b)$
Lower: $B - t(s_b)$

Confidence intervals can be calculated in a similar way.

For some types of data, such as opinion surveys, which have employed Likert scales or something similar, the data are not suitable for analysis by regression and correlation due to the nature of the scales employed.

Example
Data on satisfaction with projects procured by various approaches have been collected from clients and contractors using a 5-point Likert scale (1 = totally satisfied, 5 = totally dissatisfied). The data are summarised below. All cell figures are numbers of respondents scoring per cell.

Procurement method	Client Satisfaction					Contractor Satisifaction				
	1	2	3	4	5	1	2	3	4	5
Traditional	2	4	7	6	3	1	5	10	0	4
Construction management	3	5	1	8	2	5	5	5	3	2
Management contracting	4	6	2	8	0	6	6	5	2	1
Project management	2	4	6	2	6	3	7	3	4	3
Design & build	5	2	6	3	4	6	6	5	1	2
Design & management	2	3	7	4	4	5	5	3	5	2
BOOT*	1	6	7	3	3	6	6	4	2	2

*The procurement method: Build, Own, Operate, Transfer.

Means and standard deviations are not appropriate. Rank correlations should be used as only the rankings can be compared. Rankings of satisfaction are obtained from examining the scorings in the totally satisfied column (column 1) – only in the event of tied scores are the next (and then subsequent) columns considered to determine the ranks.

	Clients	Contractors	D (Difference)	D^2
Traditional	4	7	3	9
Construction management	3	4	1	1
Management contracting	2	1	1	1
Project management	5	6	1	1
Design & build	1	2	1	1
Design & management	6	5	1	1
BOOT	7	3	4	16
				30

The approach is adopted because respondents interpret the scorings differently in terms of both levels of the score categories (1 to 5 and associated word descriptors, if any) and of the intervals between the score categories. The latter differences may be accentuated if the data are collected against word descriptors only – semantic differentials.

The coefficient of correlation between the ranks is a measure of the association between two variables which is determined from the ranks of observations of the variables. It is calculated using Spearman's coefficient of rank correlation, ρ:

$$\rho = 1 - \frac{6 \sum D^2}{n(n^2 - 1)} = 1 - \frac{6 \times 30}{7(7^2 - 1)}$$

$$= 1 - \frac{180}{7 \times 48} = 1 - 0.54 = \underline{0.46}$$

As Likert scales yield ordinal data, strictly, regression and correlation cannot be used, as those analytic techniques require interval data. However, following Labovitz (1970), especially where ordinal variables permit a large number of categories to be specified, the variables can be treated as interval data, especially as techniques like regression and correlation are well known, powerful and quite easy to use and interpret. The view is controversial and so, before treating ordinal data as interval data, advice should be obtained form an expert statistician regarding the validity of such an adaptation for the particular data: if in doubt, be strict in the treatment of data.

Time series

Many of the data which are used by researchers are time series. These are measurements of a variable, such as temperature of the air at a particular location, made at constant intervals of time over a period. As the measurements are instantaneous representations of a variable which changes continuously, joining the points on a plot produces a graph of the time series.

Time series have four component parts:

- Secular trend (T)
- Cyclical fluctuation (C)
- Seasonal variation (S)
- Residual component irregular/random variation (R)).

Quite simple techniques can be employed to break down the time series into its deterministic components; however, certain aspects must be considered first. The *nature* of the model – the way in which the components aggregate to produce the realisations – is either *additive* or *multiplicative* (i.e. $A = T + C + S + R$; or $A = T \times C \times S \times R$). The relevance of the data must be evaluated – annual data cannot reveal a seasonal component, and short runs of data cannot reveal cycles. In economics, cycles are short, medium and long; long cycles may be of around 50 years.

If the data cannot reveal certain components, the *hidden* component is considered to be a joint part of the residual.

Having collected realisations of, if possible, raw data, the first step is to produce a scatter plot, then the line of best fit is used to represent the secular trend. Often, line fitting techniques, using the regression techniques described earlier, are used to determine the line of best fit. The usual criterion for determining the line of best fit is minimum least squares error. For long duration runs of data, it is important not to fit a secular trend line of too complex a shape such that it will absorb some seasonal/cyclical components. It is best to select a line of quite simple shape and well-known mathematical formula – straight line, logarithmic exponential, Gompertz, logistic, polynomial. A good criterion is to use the simplest, appropriate form of line; this approach is called 'parsimony'.

The standard mathematical equations for the lines just noted are:

straight:	$y = a + bx$
logarithmic:	$y = ab^x$
exponential:	$y = ae^x$ (a 'special case' of logarithmic)
Gompertz:	$y = ka^b$
logistic:	$1/y = k + ab^x$
polynomial:	$y = a + bx + cx^2$; $a + bx + cx^2 + dx^3$ etc.

Familiarity with the shapes of the lines produced by the standard form equations will be a notable help in trend-fitting. It is usually better to avoid high powers of x.

Alternative methods to mathematical line fitting to represent the secular trend include *semi-averages* and *moving averages*. For semi-averages, the data set is divided into halves, the average of each half-set is calculated and the semi-average trend line is the straight line through the two half-set average points, i.e. the semi-average points. For moving averages (MA), an appropriate number of data points must be selected for averaging. The selection depends on the nature of the data, the periods of sampling, and the amount of 'smoothing' desired, since the more the data points are averaged, the greater the smoothing. Using an odd number of data points, the moving average is 'centred' on a sampling time automatically. However, for a moving average with an even number of data points, weightings must be used to 'centre' the averages on sampling times.

Example
Assume quarterly data; i.e. sampling is at 3-monthly intervals. In this instance, averaging is over three quarters.

Quarter	Data	Total for 3 quarters	3QMA
1	10		
2	11		11.3
3	13	34	12.7
4	14	38	13.7
5	14	41	14.7
6	16	44	15.0
7	15	45	16.3
8	18	49	

To calculate the 4QMA (common for quarterly data) from the last table of data, the formula to centre the moving averages on sampling times is:

$$\frac{A_{t-2} + 2A_{t-1} + 2A_t + 2A_{t+1} + A_{t+2}}{8}$$

A = actual data point
A_t = actual data value at time t

The moving average is centred on A_t with the preceding and succeeding quarterly data at distance 2 quarters given a single weighting each, other data points being given a double weighting each, to preserve a balanced centring.

Example

Quarter	Data	Weighted 4Q total	4QMA
1	10		
2	11		
3	13		12.5
4	14		13.62
5	14	100	14.5
6	16	109	15.25
7	15	116	
8	18	122	

In order to fit a straight line ($y = a + bx$) to a set of sample data:

$$a = \bar{Y} - b\bar{X} \quad \text{and} \quad b = \frac{\sum XY - n\bar{X}\bar{Y}}{\sum X^2 - n\bar{X}^2}.$$

For the sample data, 'rounded' about the mid point of X, i.e. $\bar{x} = 0$ (see table), this reduces to $a = \bar{y}$ and $b = \sum xy / \sum x^2$.

Year/ Qtr	(Data; A) Output – Const 1975 prices [y]	Round Q1 1977	Round Q1 1978/ Q4 1979 [x]	x^2	xy	Linear Trend (T) $y_t = 2745 + 7.28x$
1977 1	2511	0	−15	225	−37665	2636
2	2612	2	−13	169	−33956	2650
3	2698	4	−11	121	−29678	2665
4	2654	6	−9	81	−23886	2679
1978 1	2562	8	−7	49	−17934	2694
2	2854	10	−5	25	−14270	2709
3	2910	12	−3	9	−8730	2723
4	2804	14	−1	1	−2804	2738
1979 1	2523	16	1	1	2523	2752
2	2848	18	3	9	8544	2767
3	2951	20	5	25	14755	2781
4	2908	22	7	49	20356	2796
1980 1	2764	24	9	81	24876	2811
2	2809	26	11	121	30899	2825
3	2858	28	13	169	37154	2840
4	2648	30	15	225	39720	2854
	43914		0		9904	

The best prediction of the series shown in the last data column is the trend plus seasonal components. The sample data were insufficient to yield a cyclical component and the residuals are random. Given appropriate analysis of the other components, over time, the mean of the residuals is zero.

Note: in the example, the additive model where $A = T + C + S + R$, was used; in practice (especially in economics), the multiplicative model is preferable. No cyclical fluctuation can be calculated in this example as four years is not sufficient data, hence the cyclical fluctuation component is absorbed by each of the three other components. Hence, in the table below, the residual component is calculated as $R = A - T - S$.

Year/ Qtr		Linear trend (T)	Detrended series (A − T)	Seasonal component (S)	Seasonally adjusted series (A − S)	Residual component (R)	Predicted series
1977	1	2636	−125	−121	2632	−4	
	2	2650	−38	39	2573	−77	
	3	2665	33	98	2600	−65	
	4	2679	−25	−17	2671	−8	
1978	1	2694	−132	−121	2683	-11	
	2	2709	145	39	2815	106	
	3	2723	187	98	2812	89	
	4	2738	66	−17	2821	83	
1979	1	2752	−163	−121	2644	−108	
	2	2767	81	39	2809	42	
	3	2781	170	98	2853	72	
	4	2796	112	−17	2891	129	
1980	1	2811	−47	−121	2885	74	
	2	2825	−16	39	2770	−55	
	3	2840	18	98	2760	−80	
	4	2854	−206	−17	2665	−153	
1981	1	2869		−121			2748
	2	2883		39			2922
	3	2898		98			2996
	4	2912		−17			2895
1982	1	2927		−121			2806
	2	2842		39			2981

The detrended series is used to construct a table, as shown below, so that the seasonal components of the time series can be calculated.

Year	IQ	2Q	3Q	4Q	
1977	−125	−38	33	−25	
1978	−132	145	187	66	
1979	−163	81	170	112	
1980	−47	−16	18	−206	
TOTAL	−467	172	408	−53	[60]
CORRECTED:	−482	157	393	−68	
[60/4 = 15]					
AVERAGE [÷4]	−121	39	98	−17	[To nearest whole number]

Index numbers

Index numbers are a means of measuring changes of a *composite* entity over time which, itself, is not quantifiable directly, but the components of which may be measured and aggregated. Common examples of index numbers are the retail prices index and the FT100 share index. The composite nature of an index number distinguishes it from a *price* or *quantity relative*. A quantity relative is the quantity of an item in the current year compared to the quantity in the base year (e.g. quantity bought in these years described as a number of units bought). A price relative is the price of an item in the current year compared to the price in the base year (usually price per unit). Index numbers use the concept of a 'basket of goods', which is a composite entity, to provide the weightings to be applied to price relatives.

Example
Price relative.

Price Relative

Year	Price/unit (£)	Index
1950 (base year)	20.00	100
1955	24.00	120
1960	29.00	145
1965	34.00	170
1970	40.00	200
1975	47.00	235
1980	58.00	290
1985	70.00	350

Index numbers are used extensively in published statistics. They show the change from a base point of index 100. Analysis of changes between points, other than with the base, requires an element of arithmetic. A further element of possible difficulty or confusion is that if two series of index numbers, say a price index – PI – and a cost index – CI, have the same base point and both are allocated 100 at that point, e.g. 1 January 1990, it does *not* mean that prices and costs were equal at 1 January 1990; subsequent proportional changes in the two index series can be considered.

Example

If at 1 April 1991, PI = 120 and CI = 115, prices have increased, on average, 20% over the 15 month period whilst costs have increased by 15%. Thus, over the 15 month period, the activity became more profitable – by 5% of the base price.

Simple average index

The simple average index takes no account of the different units of sale, patterns of consumption and changes in those patterns.

Example

Item	Price/unit 1960 (pence)	Price/unit 1985 (pence)
Bricks	20	40
Electric cable	35	60
Sand	25	50
PVC pipe	30	40
Cement	60	100
Aggregate	20	30

$$I = \frac{40 + 60 + 50 + 40 + 100 + 30}{20 + 35 + 25 + 30 + 60 + 20} \times \frac{100}{1}$$

$$= \frac{32\,000}{190}$$

$$= \underline{\underline{168}}$$

Production of index numbers comprises relating ratios which express the change in the price or quantity to a particular parameter – usually time. Price relatives are expressed for a pre-determined quantity. Normally, determination of price relatives involves sampling; the average price relative is used for the index. The variability of the price relatives for the goods in the basket about the average is important in establishing and maintaining the validity of the index.

Components of an index:

$$\text{Prices in year} \quad \begin{array}{c} \\ 0 \\ 1 \\ 2 \end{array} \begin{bmatrix} \sum p_0 q_0 & \sum p_0 q_1 & \sum p_0 q_2 \\ \sum p_1 q_0 & \sum p_1 q_1 & \sum p_1 q_2 \\ \sum p_2 q_0 & \sum p_2 q_1 & \sum p_2 q_2 \end{bmatrix}$$

$$\begin{array}{c} \text{Quantities in year} \\ 0 \qquad 1 \qquad 2 \end{array}$$

$p_0, p_1, p_2, =$ prices in years 0, 1 and 2

$q_0, q_1, q_2, =$ quantities in years 0, 1 and 2

The main diagonal, or trace, of the matrix (top left to bottom right) gives measures of value of the constituents of the index.

$$V_{01} = \frac{\sum p_1 q_1}{\sum p_0 q_0} \qquad V_{02} = \frac{\sum p_2 q_2}{\sum p_0 q_0}$$

$V_{01} =$ change of value from year 0 to year 1

The relationship for the change in value from year 1 to year 2 is:

$$V_{12} = \frac{\sum p_2 q_2}{\sum p_1 q_1}$$

If the index comprises a variety of items, i, where $i = 1, 2, 3, \ldots, n$, then

$$\sum_{i=1}^{n} p_{i0} q_{i0} = \sum p_0 q_0$$

A 2-case matrix:

$$\text{Prices} \quad \begin{array}{c} 0 \\ 1 \end{array} \begin{bmatrix} \sum p_0 q_0 & \sum p_0 q_1 \\ \sum p_1 q_0 & \sum p_1 q_1 \end{bmatrix}$$

$$\begin{array}{c} \text{Quantities} \\ 0 \qquad 1 \end{array}$$

Base weighted and *current weighted* indices are standard forms of index numbers.

The *base weighted* (Laspèyres) indices are:

$$\text{Price:} \quad \frac{\sum p_1 q_0}{\sum p_0 q_0} \quad \text{i.e. } P_{01}(q_0)$$

$$\text{Quantity:} \quad \frac{\sum p_0 q_1}{\sum p_0 q_0} \quad \text{i.e. } Q_{01}(p_0)$$

Laspèyres indices are base weighted (usually at year 0). As the weighting applies to the run of index numbers thereafter, the weightings are *fixed* (as at the base).

The *current weighted* (Paasche) indexes of price and quantity are:

$$\text{Price:} \quad \frac{\sum p_1 q_1}{\sum p_0 q_1} \quad \text{i.e. } P_{01}(q_1)$$

$$\text{Quantity:} \quad \frac{\sum p_1 q_1}{\sum p_1 q_0} \quad \text{i.e. } Q_{01}(p_1)$$

The first row of the matrix denotes expenditure at constant prices at the base year; columns represent expenditures at constant quantities (of the particular years). Quantity indexes represent expenditures in 'real terms'.

Laspèyres indexes are both *fixed weighted* and *base weighted*. It is usual for the reference base (year) to be the year used for the weights base. Most published indexes are rebased periodically, this helps to maintain their validity.

A Laspèyres price index, $P_{0t}(q_0)$ can be re-based whilst retaining its fixed weighting by switching rows:

$$P_{1t}(q_0) = \frac{P_{0t}(q_0)}{P_{01}(q_0)} = \frac{\sum p_t q_0}{\sum p_1 q_0}$$

It can retain its base weighting but obtain a different reference base by switching columns:

$$P_{1t}(q_t) = \frac{\sum p_t q_1}{\sum p_1 q_1}$$

The construction of some simple index numbers occurs as follows:

Example

Weekly wage rates (£)

Employee type	1950	1955	1960
Unskilled	8.75	10.75	14.50
Semi-skilled	9.75	12.50	16.00
Skilled	12.00	14.00	18.50
Clerical	10.00	12.25	13.00
Total	**40.50**	**49.50**	**62.00**

Weekly wage rate relatives, 1950 = 100

Employee type	1950	1955	1960
Unskilled	100	122.9	165.7
Semi-skilled	100	128.2	164.1
Skilled	100	116.7	154.2
Clerical	100	122.5	130.0
Total	**400**	**490.3**	**614.0**

Index of weekly wage rates:

$$1950: \quad \frac{400}{4} = 100.0$$

$$1955: \quad \frac{490.3}{4} = 122.6$$

$$1960: \quad \frac{614.0}{4} = 153.5$$

By using aggregated wage rates:

Aggregates: $1950 = 40.50$; $1955 = 49.50$; $1960 = 62.00$

$$1950: \quad \left(\frac{40.5}{4} \div \frac{40.5}{4}\right) \times \frac{100}{1} = \frac{40.5}{40.5} \times \frac{100}{1} = 100$$

$$1955: \quad \frac{49.50}{40.50} \times \frac{100}{1} = 122.2$$

$$1960: \quad \frac{62.00}{40.50} \times \frac{100}{1} = 153.1$$

	(Quantity weights) No of employees	(Value weights 1950) Average wages bill (£)
Assume constant 1950–1960		
Unskilled	9	102.00
Semi-skilled	23	293.25
Skilled	17	252.17
Clerical	1	11.75
Total	**50**	**659.17**

Relatives	Average wages bill (1)	Wage rates relative to: 1950 (2)	1955 (3)	1960 (4)	(1) × (2)	(1) × (3)	(1) × (4)
Unskilled	102.00	100	122.9	165.7	10200	12535.800	16901.40
Semi-skilled	293.25	100	128.2	164.2	29325	37594.650	48151.65
Skilled	252.17	100	116.7	154.2	25217	29428.239	38884.63
Clerical	11.75	100	122.5	130.0	1175	1439.375	1527.50
Total	**659.17**				**65917**	**80998.064**	**105465.18**

$$1950: \quad \frac{65\,917}{65\,917} = 100.00$$

$$1955: \quad \frac{80\,998.064}{65\,917} = 122.9$$

$$1960: \quad \frac{105\,465.18}{65\,917} = 160.0$$

Using numbers of employees in each category and their wage rates yields index numbers of:

$$1950: \quad 100.0$$

$$1955: \quad 122.7$$

$$1960: \quad 159.8$$

Hence, the method of producing the indexes does (if only marginally) affect the result.

Frequently, long series of index numbers have the base changed periodically; the manipulation required to convert to a common base is quite straight-forward.

Example

Consider the following:

Year	Index (1955 = 100)
1960	119.0
1961	121.3
1962	125.0
1963	126.4
1964	128.0
1965	132.0
1965	100.0 (1965 = 100)
1966	102.1
1967	104.0

To convert the second run to 1955 = 100 base:

$$1965: \quad 132.0 \quad \times \quad \frac{100.0}{100.0} \quad = \quad 132.0$$

$$1966: \quad 132.0 \quad \times \quad \frac{102.1}{100.0} \quad = \quad 134.8$$

$$1967: \quad 132.0 \quad \times \quad \frac{104.0}{100.0} \quad = \quad 137.3$$

Chained index

Often, short runs of index numbers are produced which must be 'spliced', i.e. joined together coherently to produce long runs. This is similar to the change of base calculation.

An alternative is to produce a chain index. Although, ideally, a chain index would be updated continuously for changes in quantities as well as prices, in practice, for indexes such as the Retail Prices Index (RPI), *chaining* i.e. adjusting quantities etc., occurs annually.

Example

Year	GDP at 1958 constant prices (£m)	Index	Price Index 1958 = 100
1956	21070	$\frac{21070}{21478} \times \frac{100}{1} =$	98.1

Year	GDP at 1958 constant prices (£m)		Index	Price Index 1958 = 100
1957	21474	$\dfrac{21474}{21070} \times \dfrac{100}{1} =$	101.9	100.0
1958	21478	$\dfrac{21478}{21474} \times \dfrac{100}{1} =$	100.0	100.0
1959	22365		104.1	104.1
1960	23484		105.0	109.3
1961	24268		103.3	113.0
1962	24442		100.7	113.8
1963	25537		104.5	118.9
1964	26912		105.4	125.3

Index numbers are a convenient way of representing a time series to demonstrate relative, proportional changes from the base. Beware of changes occasioned by sampling variations, splicing, chaining etc. Remember that index numbers do *not* give absolute figures.

As it is common for index numbers to be used to depict time series data, it is appropriate to reinforce the wisdom by drawing a diagram of the data in as 'raw' a state as possible. An important aspect of time series is that the realisations are subject to a constant set of probability laws – this is a vital consideration for predictability; an essential of management. Plotting the data makes it easy to identify the likelihood of the probability requirements being met and, often more obviously, any incidences of 'discontinuities' or *shocks* – external (*exogenous*) influences which cause a disruption to the 'smooth' flow of the data stream.

Shocks are important but, due to their nature, cannot be predicted. In his major work, 'Building Cycles and Britain's Growth', Parry Lewis (1965) considered the most important influences on levels of construction activity to be:

- population,
- interest rates, and
- shocks.

Summary

This chapter has considered several techniques for the analysis of data. It is useful to aim to maintain simplicity; understanding what analyses are being undertaken and why, and their validity, is *paramount*. Usually, it is helpful to plot the raw data to gain a first impression of any patterns. We have examined the basics of regression and correlation, time series and index numbers. It is important to use only those tests and techniques which are appropriate to the data, so awareness of the nature of the data collected is vital. Often, results in the forms of hierarchies will be obtained – in such instances, rank correlations may be useful. In cases of complexity, it is likely to be a good idea to consult the tutor, supervisor or/and a statistician.

References

Geddes, P. (1968) *Cities in Evolution*, Benn, London.

Labovitz, S. (1970) The assignment of numbers to rank order categories, *American Sociological Review*, 35, 515–524.

Levin, R.I. & Rubin, D.S. (1990) *Statistics for Management*, 5th edn, Prentice-Hall,

Lingoes, J.C. (1968) The multivariate analysis of qualitative data, *Multivariate Behavioural Research*, 3, 61–64.

Moser, C.A. & Kalton, G. (1971) *Survey Methods in Social Investigation*, 2nd edn, Dartmouth.

Parry-Lewis, J. (1965) *Building Cycles and Britain's Growth*, Macmillan, London.

Pindyck, R.S. and Rubinfeld, D.L. (1981) *Econometric Models and Economic Forecasts (2nd edn)*, McGraw-Hill.

Popper, K.R. (1989) *Conjectures and Refutations: The Growth of Scientific Knowledge*, Routledge & Kegan Paul,

Wilson, M. (1995) Structuring qualitative data: multidimensional scalogram analysis. In: *Research Methods in Psychology*, (eds Breakwell, G.M., Hammond, S. & Fife-Shaw, C.), pp. 259–275, Sage, London.

Yeomans, K.A. (1968) *Statistics for the Social Scientist: 2 Applied Statistics*, Penguin Books Ltd, Harmondsworth.

Part 3

Reporting the Results

Chapter 8

Results, Inferences and Conclusions

> The objectives of this chapter are to:
>
> - discuss the **requirements for valid results**;
> - consider alternative ways of producing and presenting **results**;
> - examine the use of statistical **inferences**;
> - examine the requirements of **conclusions**.

Requirements for valid results

Once the research project has been structured, the theory and literature studied, the data collected and analysed, the next stages are to
produce results and, by making inferences, to draw conclusions and
make recommendations. The results relate to the analyses of data,
whilst the conclusions use those results, together with the theory
and literature, to determine what has been found out through the
execution of the study. Particularly, conclusions must relate to any
hypotheses proposed, the objectives and, hence, to the overall aim
of the research.

It is important to be sure of the validity of the work – the confidence which someone may have in the findings. One should judge
how the findings may be used in other research and in application in
practice. Part of such appreciation leads to recommendations for

further research – this is identification of additional areas of study to extend and complement the work which has been carried out; it will inform the development of subsequent research projects. Thus, results are what emerge from analyses and, as such, require interpretation to give meaning in the context of what the research sought to discover. They must demonstrate what has been found out through the execution of the research.

For quantitative studies, statistical inference is employed to determine the applicability of the results to the issues under investigation and, thence, the drawing of conclusions.

Example

In the production of concrete, slump tests are employed to test the conformity of the batch of concrete with the specified mix for workability and strength requirements. If the results lie within the limits prescribed, it will be inferred that the concrete will be suitable, and vice-versa. Further, cube strength tests for concrete at 7 and/or 28 day strengths use comparisons of test results with design requirements over time to make inferences about the strength of the concrete tested.

Note, in both tests, the importance of following the correct, prescribed test procedures for inferences to be valid. Using a poker vibrator on site to vibrate a 150 mm concrete test cube may give results of very high strength, but those results are utterly useless for making inferences about the strength of the concrete placed on site because of the radically incorrect method of carrying out the test.

Conclusions take a 'broad perspective', looking at the research executed as a whole, but focusing particularly on the hypotheses, objectives and aims of the research, adopting an incremental approach to generalisations which may be made.

Almost inevitably, important issues will be identified during the course of the research – some will be incorporated into the study whilst others will remain outside its scope. The issues incorporated should be subject to conclusions, whilst those identified but not researched should be noted in the recommendations for further research.

Results

Producing the results

Results record the outcomes from tests. The selection of appropriate tests to analyse data is very important. In some cases, a variety of analyses may be employed, both statistical and/or non-statistical. Given sufficient time and other resources, it is useful to employ 'triangulation', using a variety of analyses of the data so that the results which are produced can be considered both from the viewpoints of the individual analyses and from the perspective of the combination of the analyses. In particular, attention can focus on the analyses which produce results which agree with each other broadly, if not exactly, and any which produce conflicting results. Not only are the results dependent on the tests which have been carried out, but are dependent upon the data which have been collected and the recording of those data and their coding, if applicable.

Naturally, for some 'practical' purposes, small errors in coding data may not be significant, but for any type of research, errors are likely to be material. Thus, elimination of errors requires both careful design of the coding system to ensure clarity and ease of allocation, plus checking of the allocations. Apart from mistakes in allocations, ambiguity is the main source of errors. Such misallocations will, at least, distort the results.

Triangulation may be carried out by collecting several sets of data. Often the data will be collected from different samples, although sub-sets of primary samples may be used – such as where interviews are conducted with a sub-set of respondents to a questionnaire survey sample. In such circumstances, it is important that an appropriate, unbiased method is used to select the sub-set. The sub-set should be selected either by random sampling or by following the sampling frame. Naturally, this will be limited by those respondents who do not agree to participate in interviews and so the issue of non-responses must be addressed.

Introductory results

Commonly, research projects yield a considerable number of individual results from the analyses of the data. Some results will be simple – in most cases, introductory results will be needed to provide a suitable structure for the data employed for the analyses.

Assuming that the sampling frame, methods and samples sought have been noted, introductory results to describe the data analysed for each set and sub-set should include:

- useable response numbers and rates (expressed as percentages)
- descriptive statistics *relevant to the investigation* – e.g. sizes of organisations, mean turnover, standard deviation of turnover
- description of individuals responding – e.g. job title (to indicate nature of views provided and credibility and authority of those views).

Such results would complement the discussion of how the data were sought by describing the actual sample analysed compared with the sample desired. For laboratory experiments, the experimental technique, equipment etc. will be described and discussed, whilst for modelling and simulations the sources and nature of the data and information and the methods for building the model or conducting the simulation will be noted.

Minor differences between samples desired and those obtained are unlikely to be of great importance. However, if samples sought are small and, given that tests on smaller samples are usually less reliable than tests on larger samples, small numbers in the samples will mean quite large percentage differences - it is the proportionality of differences which tends to be important. Thus, small numerical differences may cause samples available for testing to be so small that any results may be (almost) meaningless.

The second aspect to examine is the pattern of the differences. If the differences are random, or follow the sampling frame proportionately, they are of less significance than if they follow a particular, different pattern. Where this occurs, and the differences show bias in themselves, they introduce bias into the sample analysed and, hence, the results obtained. So, differences between samples desired and those obtained may affect the validity of the results, at least in terms of confidence, due to size of data sets useable and considerations of bias.

It is useful to determine why non-responses have occurred and, if possible, to investigate whether groups of non-respondents differ in their views etc. from the respondents and from other groups of non-respondents. For low response rates and small response numbers, the views of non-respondents will be dominant but, without appropriate investigation, unknown and the validity of the research findings is questionable. In this situation it may be advisable to constrain the sample and its size, and to review the data collection method such that response rates from the revised sample will be sufficient to ensure only small possible non-response bias. Apart from the issue of non-response bias, significantly lower response rates than those anticipated may mean that statistical testing may require modification from tests valid on large samples to those applicable to small samples. The boundary between 'large number' and 'small number' statistics is at $n = 32$ although the size adopted in practice often is $n = 30$ (see Levin & Rubin, 1991), where n is the sample size.

Substantive results

Having provided a description of how to subject the data to analyses to yield the results, the substantive results themselves may be considered. For experiments, modelling and simulation, sources of possible errors should be noted, and as far as possible quantified, to yield an overall measure of 'experimental error' (i.e. a variability/reliability measure). Although certain results will be sufficient in themselves to provide the required information, others will need aggregation, companions and rearrangement to enable researchers to maximise understanding of what the results mean. Different forms of presentations will be useful also – notably charts and graphs; tables may convey precision but are much more difficult to interpret.

Especially for applied research projects, in which objectives will have been set at the outset, and where the research is targeted towards the provision of answers to particular questions or issues, the results, and the research leading to them, can be categorised as follows (following Ritchie and Spencer, 1994):

- Contextual: identifying the form and nature of what exists
- Diagnostic: examining the reasons for, or causes of, what exists
- Evaluative: appraising the effectiveness of what exists
- Strategic: identifying new theories, policies, plans or actions.

Clearly, there should be a fairly 'seamless' transition from results of analyses to the conclusions drawn. Often that transition leads to lack of appreciation of the differences between results and conclusions. The differences, however, are quite marked – results are produced from analyses of data, whilst the conclusions consider the results in the contexts of the topic and the whole of the study in order to determine what exists, why and how, and the considered consequences of issues examined in the research.

For projects which obtain qualitative data, the analysis will initially have involved the researcher's getting 'immersed' in the data to gain maximum familiarisation; the first of the five analytic stages noted by Ritchie and Spencer (1994). The second stage is identification of a thematic framework followed by indexing, charting and, finally, mapping and interpretation. At this last stage the key objectives of the analysis are addressed. In such studies, the 'development' of results via analyses is more appropriate a view than the 'production' of results from analysis, which is appropriate in quantitative studies.

The stages of charting and mapping and interpretation involve the production of results. The charts include headings and sub-headings and are laid out 'thematically' – by themes across responses, or by case – for respondents across the themes. Mapping uses the charts to extract key characteristics from the data. Whilst the charts and analytic processes leading to their production present the raw results; mapping enables the researcher to juxtapose alternative views, to set out pros and cons such that the charts present evidence which can be structured through mapping.

Inferences

Inference is the process by which the meanings and implications of the results are determined in order that conclusions may be drawn. To assist the process of inference, the form and levels of aggregation of the results must be considered. Commonly, the transition from results to conclusions of a research project is effected in the report by a section or chapter involving discussion of the results. These will be discussed in the contexts of each other and of the theory and literature. Throughout, it is appropriate to ensure that the discussion is relevant to the pursuance of the aim and objectives of the work although it is usual for other issues to emerge.

In order to gain full understanding of the results and appreciation of their implications, it is helpful to present the results in a variety of ways, including tables and diagrams. The diagrams provide immediate indications of generalities, trends etc., whilst the precision of detail contained in tables is important to facilitate particular considerations and subtleties.

In considering applied policy research using, particularly, qualitative data, Ritchie and Spencer (1994) note that, in seeking associations, '...the analyst may become aware of a patterning of responses...it may appear that people with certain characteristics or experiences also hold particular views or behave in particular ways...the analyst will systematically check for associations between attitudes, behaviours, motivations etc....'. Analogous considerations apply to quantitative results.

By inference and discussion, researchers seek to gain insights into how the individual parts of the study relate to each other in terms of the issues which the individual parts identify. In the absence of theory, results are isolated; their consideration in the light of theory and previous research findings facilitates the advance of knowledge and a perception of how the topic and its practice is developing.

Causal relationships

Results must be interpreted in the context of each other, of the theory and of results of previous research. Such interpretation is intended to reveal relationships between the results in terms of extent and, it is hoped, direction, as well as helping to gain insights into causation.

A difficulty in investigating causal relationships between pairs of variables is the potential causal influence of other variables. Moser and Kalton (1971) consider how the effects of such further variables can be dealt with in the evaluation using multi-variate analysis techniques. A more fundamental notion is to employ theory to examine the logic of causation.

> *Example*
> There may be a correlation between phases of the moon and, lagged by the appropriate time interval, births of babies; such correlation does *not* demonstrate causality and what knowledge is available indicates very strongly that it is something other than the phases of the moon which causes births of babies.

In some instances, although theory may suggest direction and causation of relationships between variables, it may not be conclusive. This is quite common in the social sciences; for example, it is not clear whether demand leads supply or vice versa, and perhaps both relationships operate in the dynamics of a market economy. Further, a variety of theories may be appropriate – such as those relating to inflation, for Keynesian, Monetarist theories etc.

Hence, the discussion of results in the contexts of theory allows alternative considerations and views to be examined. Results should indicate strengths of relationships – often epitomised by the confidence level, determined by the degree of statistical significance. They should also provide the numerical statistical measure of the relationship, taking sample size into account. Thus, inference requires an holistic view of the research project to be adopted.

In studies which incorporate both qualitative and quantitative data and analyses, it is essential to bear in mind that the data are founded on different logical principles and, consequently, so are the results. Therefore, it is important to be aware of both the principles and the sets of assumptions, which should be explicit, in order to assist the researcher to move between the two sets of results with ease.

Interpretation

Inferences and discussions enable the researcher to present the issues arising out of the research from two perspectives separated in time – that prior to the execution of the empirical work and that following its execution and production of results. Comparison of the two perspectives is important to demonstrate how knowledge has changed due to the study – to reinforce or to question the previously 'perceived wisdom'.

In interpreting results, associations and causalities between variables are investigated. Usually, variables are considered in pairs; independent and dependent variables. In doing so, 'third' variables must be eliminated by the sampling approach adopted or by adjustment in the analysis, as demonstrated in the following example.

Example

A random sample of 2000 people who smoke is selected to investigate whether a short TV campaign will induce them to give up smoking. Six weeks after the end of the campaign, they are asked whether they have given up smoking:

	Viewed (V)	Not Viewed (N)	Total
Still Smoking (S)	500	300	800
Given up (G)	1000	200	1200
	1500	500	2000

67% of those who viewed the campaign had given up smoking whilst only 40% of those who did not view the campaign had given up. The results suggest that the campaign was successful but other factors (variables) could have been influential – consider 'social class' and 'age group'.

	(V)	(N)	Middle Class (M) (V)	(N)	Working Class (W) (V)	(N)
(S)	500	300	125	75	375	225
(G)	1000	200	250	50	750	150
	1500	500	375	125	1125	375

Analysis by 'elaboration' of the 'social classes' variable yields the same percentage of viewers and non-viewers who gave up smoking in both of the social classes as in the total sample. Hence, 'social class' is not a significant variable.

	(V)	(N)	Younger (Y) (V)	(N)	Older (O) (V)	(N)
(S)	500	300	275	225	225	75
(G)	1000	200	475	25	525	175
	1500	500	750	250	750	250

> In the older age group, 70% of those who viewed and 70% who did not view gave up smoking, whilst in the younger group 63% of those who viewed the campaign gave up smoking whilst only 10% of the non-viewers in that age group gave up.
>
>   Clearly, age group is significant, irrespective of the TV campaign, for older people, whilst the campaign was significant for the younger people in inducing them to give up smoking.

Associations *within* groups are *partial* associations and *between* groups are *marginal* associations. For the social class analysis, marginal associations of social class with viewing the TV campaign and with giving up smoking do not explain any part of the overall association; the association is explained by the partial association completely. This is termed *elaboration by partials*. For the age group analysis, marginal and partial associations are apparent, and so both types of elaboration are relevant.

Conclusions

Conclusions are the major output of the research – what has been found out about the topic under study through the execution of the research. As such, conclusions present the major items and themes which have been raised and investigated. Conclusions complete the 'story' of the research from what it was desired to *find out* to what has been *found out*.

How to write conclusions

Generally, a research project will yield between six and twelve main points of conclusion - those facets about which the study has helped to develop knowledge, what those developments are and what they mean for practice and further research work. Ideally, each conclusion stands alone, usually as a paragraph, but with sure and clear foundations in the research which has been executed. Only if, coupled with the theory and literature, the empirical research carried out leads to the conclusion should it be stated as a conclusion; mere opinions, whims and conjectures of what could or might be are not appropriate. As grounded theory leads to the development of theories etc. by scrutiny of data collected and identification of patterns and relationships, so the conclusions of any research project must truly emerge from the research actually done; no new material should be introduced in the conclusions.

A common good practice is to produce one stand-alone, often numbered, paragraph for each point of conclusion. For studies which set out explicit aim, objectives and hypotheses, it is important that the conclusions relate back to them. Hypotheses tested through the research must be subject to conclusions. These should state whether the research leads to support or rejection of the hypotheses, with what level of confidence, why, and with what likely consequences, given the other results and conclusions. The overall aim was used to determine objectives and, thence, hypotheses through analysis. So, synthesis of the conclusions regarding the hypotheses leads to conclusions about the objectives, which in turn relate back to the aim. This approach facilitates consideration of the degree to which the objectives and aim have been realised, what further research should be done, how future studies of the topic may be executed and, by examining the validity of the study and its findings, what may be recommended for implementation and for further research.

It should be possible to reference each statement in each conclusion to a section of the research. Although such referencing should not appear in the report, it may be useful to carry out the exercise as a mechanism for arriving at the conclusions and ensuring their substantiation by the research done. In quantitative and applied studies, significant assistance in determining the topics to be addressed in the conclusions is given by the presence of a detailed aim, objectives etc. virtually from the outset. In such circumstances, those specified themes may be tracked through the work so that the major issues are addressed progressively in the theory, literature and empirical sections – this is of much assistance in developing and drawing conclusions.

Occasionally, it is useful to draw 'sub-conclusions' on completion of each section of the research project. These conclusions help to isolate what has been substantiated, what has been refuted and what remains as unresolved issues. Especially when new topics and qualitative work of investigation are being pursued, such a progressive approach can be very helpful in developing the issues to be studied as the research progresses.

In qualitative studies, the drawing of conclusions tends to be far less 'directed' due to the usual absence of detailed objectives and hypotheses; indeed, it is common for qualitative studies to set out to produce theories which are intended to inform objectives etc. for subsequent research. Hence, conclusions from qualitative studies may be in the form of identification of relevant variables, patterns and relationships between those variables. These may be either discovered by the research or by 'logical' but hypothetical issues to be the subject of future investigations. As qualitative studies tend to be concerned with research into new topics, analogies with relationships in other topics and subjects may not apply. What may be logical and rational in one discipline may not apply elsewhere, if only because different sets of assumptions are employed. Further, in moving between different groups of people, behavioural factors may alter considerably.

> *Example*
> Logical, rational behaviour in one industry may not be appropriate in others. For example, newspapers and property investment; newspapers must take a short term view in most instances whilst property investment is a long term activity.

It is essential to be aware of the variables and the assumptions as well as the principles, samples and results of analyses in drawing conclusions.

Further research

It is unlikely that any study can be comprehensive. Most hope to meet the majority if not all of the objectives, but that achievement is somewhat less than realising the aim of the research. This leads to three further facets of the research. In particular, examining the conclusions and comparing them with the aim and objectives will suggest:

- recommendations for implementation
- limitations of the study
- recommendations for further research

Evaluation of what is concluded with high levels of confidence from the study leads to any recommendations for implementation – particularly appropriate for applied research. Note what should be implemented, how and why.

Limitations explain why the scope of the study, results etc. was constrained. Certain limitations are likely to have emerged during the course of the research – not everyone who agrees to respond to a survey etc. will always do so; other circumstances change too, favourably as well as unfavourably. The limitations, especially those occurring during the research and outside the control of the researchers, are important to note with their consequences. Research is dynamic and occurs in a dynamic environment hence, change is inherent.

Recommendations for further research should suggest topics for study, appropriate methodologies, given the knowledge gained from the research just executed, and the reasons why such further studies would be useful. Mere recommendation of replicating the study with a larger sample which might reinforce the results is hardly a useful recommendation – it's obvious.

The transition from results to conclusions and recommendations through inferences requires *insight* – the conclusions should express and explain those insights such that the conclusions and recommendations will be *informative*.

Summary

The chapter has considered various approaches to the production and presentation of the results obtained from analysing the data collected. Consideration of the results in the context of the theory and literature enables inferences to be drawn and the results to be discussed in context. It is important that conclusions are drawn from the research carried out and are not mere ideas or whims. The conclusions state what has been found out during a particular research project and should relate to the aim, objectives and hypotheses, if any. Notes of the limitations of the study and recommendations for further research and for appropriate implementations should be made.

The research report presents a story in three parts:

- what the study seeks to find out,
- how the study tries to find it out,
- what the study has found out and what future studies might attempt to find out.

References

Levin, R.I. & Rubin, D.S. (1991) *Statistics for Management*, Englewood Cliffs, Prentice-Hall.

Moser, C.A. & Kalton, G. (1971) *Survey Methods in Social Investigation*, Dartmouth.

Ritchie, J. & Spencer, L. (1994) Qualitative Data Analysis for applied policy research. In *Analysing Qualitative Data* (eds A. Bryman & R.G. Burgess), pp. 173–194, Routledge, London.

Chapter 9

Reports and Presentations

The objectives of this chapter are to:

- outline the essentials for good **report production**;

- discuss the importance of effective and efficient **communication**;

- examine the **contents of the report**;

- note the essentials of **oral presentation**.

Report production

Having completed the research, it is essential to produce a report of what has been done and what has been discovered, so attention must be given to its content and form. In many instances, other forms of presentation will be required as well. It is usual for the style, layout and, sometimes, length of a research report to be specified by the course documents, the University or body commissioning the research.

The report of a research project is the primary source from which the research will inform 'the world at large' of what has been done and what has been discovered. It is a primary communication document and, hence, is of paramount importance. If the communication does not work well, the research may never realise its potential so it is essential to reflect on how to achieve good communications at the outset of producing the report. Even academic reports can be lively and stimulating if presented well, despite the possible constraints of required format, rigour and academic style. Graphics and, if viable, colour can enhance the message. It is important that the report is clear and concise; it must denote what has been done, why and how, to a level of detail to facilitate both extension of the work and its replication; an approach employed widely in experimental work.

Naturally, everyone has their own approach to writing a report. However, the report of a research project is likely to have the material assembled over a significant period of time, from several days to several years. As considered in reviewing the theory and literature, it is essential that adequate records are maintained. Some people elect to write draft sections of the report as the work proceeds which may assist distillation of the main ideas and issues; others prefer to keep notes and to write the report as an entity once the research work has been finished, the latter approach may aid continuity of themes. What is clear is that there is no universally best way to produce a report; although certain principles may be applied, the selection of the process is at the choice of the individual and so, subject to that person's preferences.

Assistance and guidance of the supervisor, if there is one, can be very significant. The approach to producing the report should be discussed and agreed with the supervisor – preferences and alternatives should be considered with the goals of securing the best report within the constraints – notably time. For researchers who are not good at producing and adhering to deadlines, it is likely to be a good idea to set and agree a schedule for the production of the report and for the supervisor to 'police' compliance with the schedule. Remember to incorporate some contingency time for revisions etc. It is a good idea to study a few research reports which in topic, level or/and purpose are similar to the one to be produced. It may be useful to do this early in the research period – at proposal formulation – to gain awareness of the academic 'target', as well as at the report production stage. An initial section or chapter draft should be submitted to the supervisor, and also to colleagues if they are willing, to ensure that what is produced is what is required; it is a good idea to submit the first portion of the report once drafted and not to write more until agreement has been reached that the portion submitted is agreed to be satisfactory.

Communication

Any communication exercise involves two parties or groups – a sender and a receiver, as shown in Fig. 9.1.

The aim is to transfer the thought of the sender as accurately and, usually, as quickly as possible to the mind of the receiver. Various languages, at least one, will be employed in the process, so common understanding of the language is essential. If translation is required, say from Cantonese to English, and neither the sender nor the receiver speaks both languages, an intermediary link in the communication chain must be introduced; the translator, through the translation process, will 'filter' the message while acting as both receiver and sender of each element of the total message being transmitted.

Clearly, noise and distortion are likely to occur to some degree in any communication; noise is interference from the environment, such as a machine operating close to where a conversation is taking place, distortion is interference in the communication system, for instance an echo on a phone line. Bias can occur in the way messages are phrased; contrast, 'Do you want to eat?' with 'Do you really want to eat?'

Feedback, an essential element of any system, to facilitate evaluation of efficiency and effectiveness, is itself a communication chain and so is subject to the potential problems of any communication

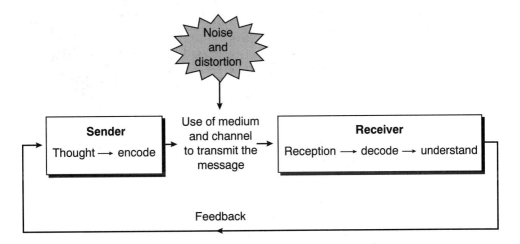

Fig. 9.1   Communication system.

system. Communication chains should be kept as short as possible, multiple forms of communication which reinforce each other in transmitting the message are useful. The most important consideration in communicating any message is the ability of the receiver to receive, decode and understand the message accurately and readily as the sender intends. If that understanding cannot be achieved, the communication has failed to a degree; the greater the difference between the intended understanding by the sender and the actual understanding by the receiver, the greater is the failure. Feedback may occur through interpretation of the actions taken by the receiver once the message has been transmitted.

The report of the research project is the primary communication to assessors, commissioners of the work, colleagues and others, of what was done, how, when and what discoveries were made. Usually, the basic form of the communication is prescribed, but other communications will occur throughout the research activities; notably with the supervisor during the production of the report.

Given the statement of requirements and limitation parameters for the report, it is vital to consider its content and how they may be presented best. It is important to write clearly, concisely, avoiding jargon and pomposity. Some writers seem to believe that it is silly to use one word when a hundred will do. Oscar Wilde apologised for writing a long letter as he did not have time to write a short one . . . leave time for reviewing the report and editing it.

Contents of the report

A research report should include:

- Title page
- Abstract
- Acknowledgements
- Contents list
- List of figures
- List of tables
- List of abbreviations
- Glossary of terms

- **Text of the report**
- References
- Bibliography
- Appendices.

How to begin

The title page may be required to follow a prescribed format. It should state the title of the research project together with the author(s), the purpose of the research, such as the partial fulfilment of the requirements for the award of a degree, and the date of completion and submission of the report.

The abstract is a very concise summary of the research report, usually in 250 to 500 words. It should outline the topic, briefly state the aim, main objectives and hypothesis, summarise the methods of data collection and analyses and state the main findings and conclusions.

The acknowledgements page is the opportunity to include brief, formal statements of thanks to people who have helped in the execution of the research – mentors, supervisors, providers of data, fellow researchers, secretaries, editors etc. Whilst it is polite to thank everyone who has assisted with the work, it is best not to be over-zealous in scope or expression and it is *most important* to ensure that confidences and anonymity assurances are not betrayed unwittingly.

The contents list, supplemented by lists of the figures, diagrams and tables, should note the chapters, main sections and sub-sections of the report, plus any appendices etc. with page numbers. The list provides an 'at a glance' overview of the report for readers; it is an invaluable guide for anyone who wishes to obtain particular facets of information without having to search the entire text.

The list of abbreviations, is a useful guide to ensure that both 'standard' and 'particular' abbreviations are understood by all readers, and that the author and readers have a common understanding. All abbreviations used in the report should be listed alphabetically with the full items noted against each abbreviation. The glossary of terms is included to ensure common understanding between author and readers. Terms which have particular meanings in the subject topics of the research and in industrial or practice context must be listed alphabetically with their definitions.

Text of the report

The text of the report constitutes the bulk of the document. Although chapter headings, sequences and breakdowns of contents will vary, notably between quantitative and qualitative research, a significant amount of contents are 'standard'. The text will begin with an introductory chapter which will discuss the topic of research, the rationale for the work, the aim and objectives. Commonly for quantitative research, the introduction will state the main hypothesis and examine the research methodology. Often, it is useful to include a diagram of variables and hypothesised relationships between them. It may be helpful to end the introductory chapter with a brief summary of the contents of the other chapters in the report and appendices, but care should be taken that the brief summaries are not just repetitions of the contents list.

Given a thorough proposal and having executed the work, planning the content in the text of the report is likely to be quite straightforward. The first and last chapters are 'Introduction' and 'Conclusions and Recommendations' *but*, in order of writing, these are the final and penultimate chapters produced, respectively. This is because the conclusions and recommendations relate to the entire work, which is outlined briefly in the introduction.

The remaining content in the text of the report is likely to be written in the chronological sequence in which the research was executed – Theory, Literature, Data Collection, Data Analysis, Production of Results, Discussion of Results. How the activities are organised into chapters is one of appropriateness.

The suggested order of writing the text of the report is:

- Theory
- Literature
- Sampling and data collection
- Presentation of results
    Data analysis
    Production of results
    Discussion of results
- Conclusions
- Recommendations
- Introduction

Theory and literature

For much research, theory is the 'bedrock' component. Although originally derived through research, it is used to inform and guide further work. All potentially, as well as actually, relevant theory should be considered and, therefore, reviewed and summarised in the report. The next element is the review of literature – the summary, juxtaposition and evaluation of the samples, methods, results and conclusions of research executed already. There must be reviews of theories pertaining to the subjects of study and a critical review of literature. The theories are basic laws, principles etc. which underpin the subjects of the research, whilst the literature is reports of researchers into developments and/or applications of theories. Such reviews may constitute one or several chapters.

Theory and literature may be combined, but they are summaries of the work of others and must be *fully and properly referenced.* The job of the researcher in producing the report is to present the salient points succinctly, demonstrating strengths of alternative views and issues, summarising what is accepted and what is subject to debate. Only from a comprehensive review should the researcher express an opinion on the weight of evidence leading to a particular line of work, view or approach being adopted. The researcher fulfils the role of an informed, thorough, objective reporter; certainly not a 'tabloid journalist'. The task in this section is to report the work of others, not to do 'a sales job'; any expressions of the researcher's view *must* be supported by the weight of evidence. The essence is to produce a critical review – the criticism must be both objective and informed. Each statement or set of statements should be referenced – opportunities for the researchers to express their own opinions will arrive soon enough.

In some studies, objectives for any empirical work, and hypothesis to be tested empirically, may not be developed until the reviews of theories and literature have been completed. In such cases, a brief chapter considering their development, with rationale, should follow the theory and literature review chapters. Particularly where the researcher is striving to formulate objectives and hypotheses, perhaps because of novelty of the topic of the study, the discipline required to produce the draft write-up of theory and literature can be invaluable in assisting objectivity and analysis of the 'state of the art' in the topic. Having carried out that exercise, it should be much easier to detect the main issues, and so to decide what aspects to research and how to proceed. Irrespective of the research approach adopted, the production of a list of issues, either within or separate from the report, is helpful to demonstrate how the particular research project fits into the debate. It also helps to advance knowledge of the subject and guides the recommendations of what further research should be done.

Almost inevitably, a researcher gets very close to, and possessive of, the research; Review by colleagues, and especially by the supervisor, is very helpful as the greater distance and consequent added objectivity ensures that what is written states what the researcher intended, and assists in the compilation of a full list of important issues. In reports of larger research projects, it is likely to be appropriate to include a separate chapter examining methodology issues and providing the rationale for the adoption of the methods used. Such a chapter may follow the introduction or, if considered more appropriate, be included between the chapters of theory and literature review and those which explain empirical work.

Reporting on sampling and data collection

The next sections of the report to be written describe and discuss the sampling; what data were collected, why, how and what the data collection involved. A discussion of the data obtained compared with the data desired is useful – if differences are minor, a note of those differences may suffice, but where differences are larger, consideration of possible and known reasons should be provided. Such a discussion should be carried forward to the consideration of results, and for examining the validity of the findings and limitations of the research. Response rates markedly different from the norms should be discussed; techniques which have produced really high response rates could be very useful to researchers in the future. Discussion of piloting, and evolution of methods due to data 'issues' should be highlighted in the report.

The main body of the report should provide the flow of the argument; those items essential to the production of the 'case'. In most instances, the report will contain a summary, including tables, diagrams and text, of the results of the useable responses. It is helpful to provide back-up detail and, for assisting future researchers, to include a copy of the questionnaire forms sent and details of the responses in the appendices. The 'summary information' provided in the main report is substantiated by the details in the appendices. Thus, the general principle is that appendices provide supplementary information or details, whilst what is included in the main report is the information and data necessary to the development of the argument – in many cases, to the testing of the hypotheses and the fulfilment of the objectives and aim.

Presentation of results

Presentation of data, tests and analyses executed, and results obtained may be in sequence or within issues such as the nature of respondents, organisational forms and descriptions etc. Non-essential details should be included as appendices. Whether the presentation is issue-by-issue or using a data-tests-results format is a matter of choice. The format which presents the information in the way in which the likely readers will assimilate and understand it most easily should be adopted. For any but small studies, the issue-by-issue approach seems to be more effective, followed by a discussion of the results which considers the compatibilities of the results – one with another – and with the contents of the reviews of theory and literature.

Generally, the tests used for analyses of data are well known. They include correlation and regression, averages, and measures of variability, rank hierarchies, measures of association. For such well-known standard tests, it is neither necessary nor desirable to include formulae and descriptions of the tests in the report – a reference to a standard text is appropriate. However, on occasions, more 'obscure' tests will be employed; it is appropriate to adopt the same approach to inclusion of their descriptions as for inclusion of the data obtained; essentials in the report and supplementaries in appendices. It is useful to discuss with the supervisor what should appear in the main report and what should be included in appendices.

Discussion of the results of empirical work requires the results to be evaluated in conjunction with the theory and literature as well as in the context of the objectives etc. In this chapter, causations should be examined and explanations weighed up. The initial model of the research variables and relationships may be modified in the light of findings of the study. Hence, to obtain appropriate detail of communication, consider what information must be included to support and demonstrate the argument and how that information should be provided to communicate the message best. Full details of the results should be provided in the appendices. The results should, as appropriate, be qualified regarding the confidence with which they can be relied upon – if possible using statistical measures, but otherwise subjectively will be helpful. To ensure a common understanding, even for quite well known terms, it is advisable to provide a glossary of terms.

Up to this stage of producing the report, the content is impersonal and does not reflect the views of the researcher. The content is facts – theory, findings from previous research, data collected, tests and analyses executed and results obtained. The discussion of results requires the researcher to evaluate the results against each other and against the contents of the theory and literature *and* to deduce explanations for similarities and consistencies and, more obviously, for differences and inconsistencies using what has or has not been done as the basis for the discussion. *If it is not in the report, including the appendices, it cannot feature in the discussion.* Clearly, the principle of including all relevant matter appropriately in the report must be followed. The discussion of the results section may reveal research which has been done but not included in the report. In this case, the report must be amended to include those items which have come to be recognised via the discussion, as important.

Conclusions

Once the results have been presented and discussed, it is appropriate to draw conclusions. Conclusions are in many ways the most important part of the report – they note the important things which have been discovered through the execution of the research. Thus, conclusions represent what has been discovered, and which can be used by other researchers and by practitioners. A concise chapter is recommended; a maximum of 2000 words is usually appropriate. Each conclusion should be a 'stand-alone' paragraph containing no new material, but firmly founded on the contents of the study. Each hypothesis merits a conclusion; whether it is supported or not, how strongly and why. The same applies to each objective, an explanation of why, and to what degree, the objective has been realised.

A conclusion should be drawn about the aim of the research. Generally, only a very small number of findings which are supplementary to the hypotheses and main objectives are adequate to merit conclusions. In research fields other than quantitative research, such as ethnographic studies, conclusions in the forms of hypotheses, objectives etc. for further study are entirely appropriate.

Recommendations

Research projects may yield conclusions which are sufficient in confidence and content to lead to recommendations for changes in practice. These are called recommendations for implementation and should be noted under an appropriate heading.

The areas in which the aim and objectives have not been met can be suggested as recommendations for further research. Further suggestions may include any potentially significant aspects which the work revealed, but which were outside the scope of the study. Each conclusion and recommendation should be included in the report as an individual, 'stand-alone' paragraph. Some studies do not produce any recommendations for implementation. The criterion is not to 'force' either conclusions or recommendations – both must be valid *from the research*; if they are not really valid, any 'forced' conclusions or recommendations will devalue the entire research.

Introduction

The final chapter to be written is the introduction. This chapter outlines the topic and the reasons for the study being undertaken. It should state the aim, objectives and hypothesis, with sub-hypotheses, and may explain the methodology, with a brief evaluation of why the selected methods have been adopted in comparison with the alternatives. In some research projects, especially where development of an appropriate methodology has been a major factor in the work, it is likely that a separate chapter is devoted to the description and discussion of the methodology adopted, and what alternatives were considered and rejected.

If possible, it is helpful to include a diagram or model of the main variables – independent, intervening and dependent – and their anticipated or hypothesised relationship. The model is derived from the aim and objectives of the study and depicts the relationships in the hypothesis or hypotheses. The model may, of course, be amended by the results of the research, and a revised vision can be shown and discussed in a later chapter in the report. The introduction should outline and introduce the total study – what is to be studied, why and how. It is *not* a summary of the work but may note the contents of chapters which follow. Hence, the logic of writing it last of all the report's chapters.

Remainder of the report

References are items of text which have been quoted or paraphrased in the report. They may quote precise extracts from texts; ensure that quotations etc. are not too lengthy and are obtained from a variety of sources to provide a comprehensive and balanced view. A bibliography is a list of texts which have been used in the research as general 'background' reading. For both references lists and bibliography lists, a standard method of referencing must be adopted, and both lists must be complete for all the source works used to support the research.

Appendices are useful supplements to the content of the main body of the report. They should be used to provide detail which might be useful for particular purposes, such as follow-up research, but are not essential to the content, understanding and 'flow' of the main report. Commonly, summaries of data collected and 'technical notes' are included as appendices.

Oral presentation

Often an oral presentation of the work by the researchers is required. Such a presentation allows the researcher to present the work and to answer questions to clarify and expand aspects of the report. On occasions, the research may be presented to an audience at a research meeting or a conference – formality will vary enormously between events. Such an opportunity can be very valuable in clarifying the main aspects of the work and in obtaining feedback and fresh views from people who are not close to the work. Such views should be objective and can be very useful in helping to determine what else should be done and by what means.

In making oral presentations, as with producing the report, it is essential to gear the presentation to the audience – to communicate the important aspects of the research in enough detail that they can be understood, but not in too much detail to bore the audience or to be confusing. It is likely that time for a presentation will be limited and, especially as most presentations are followed by a period in which the audience can question the presenter, only the main aspects of the study can be discussed. For work which is *not* complete, it is good to conclude the presentation with an outline of what the researcher intends to do to conclude the study. For 'completed' research, the presentation may end with the recommendations for further work.

As in producing a report, visual aids, projection slides etc. of graphs and diagrams are useful, but projections of major summary messages of the components used to develop the argument are very helpful too. Keep the projected messages clear and concise; one idea on each is an appropriate format. Such visual aids may note the themes of discussion, such as 'Transaction marketing'; 'Relationship marketing'; 'Price = Forecast cost plus mark-up'. These aids help to focus the attention of the audience, to reinforce the topic and to remind. Humour is great if it works but must be included with much care and sensitivity to the likely audience.

Oral presentations should not only inform, but to be used to maximum benefit should stimulate discussion, and hence feedback. It is in such situations that new lines of research can emerge (a form of 'brainstorming'). Conducting an oral presentation is likely to require the researcher to 'defend' the work; this usually involves justifying what was done and is to be done. Discussion of research from a wide variety of viewpoints is an essential of research and will demonstrate how robust the work's findings are; how well they withstand criticism and/or how generally they apply. The more rigorously the research has been executed, the more robust the findings will be.

Oral presentations provide an interactive forum in which the research can be tested, extended and improved. Most people intend their criticisms to be helpful and constructive; although it may be hard for the researcher at the time, there is great benefit to be gained from the feedback and input from oral presentations.

Summary

This final chapter has considered the communication exercise of producing and presenting the report of the research, the *dissertation* or *thesis*, and of making oral presentations. Recommendations have been made concerning both the contents and the order of preparation of the report. It is important to allow sufficient time for production, including editing and proof-reading. Get whatever help you can to ensure that the report is written objectively and that it says what you intend. It is through presentations of research projects that their worth is judged.

Index